Yours truly
Horace A. Ford

ARCHERY:

ITS

THEORY AND PRACTICE.

BY

HORACE A. FORD.

(Holder of the Champion's Medal for 1849-50-51-52-53-54-55-56-57 *and* 185

SECOND EDITION.

London:

J. BUCHANAN, 215, PICCADILLY.

CHELTENHAM: H. DAVIES, MONTPELLIER LIBRARY.

1859.

PREFACE

TO THE

FIRST EDITION.

Gentle Reader,

The favour bestowed on my late Articles upon the "Theory and Practice of Archery," published some months back in the *Field* newspaper, has induced me to present them to your notice embodied in their present more permanent form. Considerable additional matter, and the necessary illustrations and diagrams, (which were unsuited to the former mode of publication,) have now been introduced, but with this exception and some few trifling alterations, the present Work is very much a *resumè* of those Articles, a careful revision having suggested no modification of the views and theories therein laid down.

If you are already an Archer, it is hoped the perusal of the following pages may assist your onward progress in the noble Art—if one of those benighted beings who know it not, then that it may at least induce you to commence its study; having done so, there is little fear of your ever abandoning the pursuit.

Your sincere well-wisher and friend,

THE AUTHOR.

PREFACE

TO THE

SECOND EDITION.

BRETHREN OF THE BOW,

Three more National Archery Meetings have come, and gone, since the First Edition of this Work issued from the Press. The result of those Meetings has been to confirm me in my position as Holder of the Champion's Medal, and also to convince me more firmly than ever of the correctness of the theories and principles of our Art, as laid down in "The Theory and Practice of Archery." I might, perhaps, without undue vanity, lay the "flattering unction to my soul" that the great improvement which has evidently taken place during the last two years, was, in some degree, owing to the more general adoption of those principles; be this as it may, I can, at any rate, point to several of the leading Archers of the day as having attained their position in consequence of having formed, or re-formed, their practice in conformity with them. This being so, you will require no apology at my hands for introducing this Second Edition to your notice—nay, more, you will doubtless evince your appreciation of my labours in your behalf, by rendering it imperative upon me to contemplate even the possibility of a *Third* issue. Anyway, if I can but assist, in however slight a degree, in the spread and improvement of our favorite amusement, my first object, in having published my lucubrations at all, will have been gained.

Your friend,

THE AUTHOR.

CONTENTS.

CHAPTER I.

CHAPTER II.

A GLANCE AT THE CAREER OF THE ENGLISH LONG BOW.

CHAPTER III.

OF THE BOW.

CHAPTER IV.

HOW TO CHOOSE A BOW, AND HOW TO USE AND PRESERVE IT WHEN CHOSEN.

CHAPTER V.

OF THE ARROW.

CHAPTER VI.

OF THE STRING, THE BRACER, AND SHOOTING-GLOVE.

CHAPTER VII.

OF THE GREASE-BOX, TASSELL, BELT, ETC.

CHAPTER VIII.

OF BRACING AND NOCKING.

CHAPTER IX.

OF POSITION.

CHAPTER X.

OF DRAWING.

CHAPTER XI.

OF AIMING.

CHAPTER XII.

OF HOLDING AND LOOSING.

CHAPTER XIII.

OF DISTANCE SHOOTING.

CHAPTER XIV.

OF ANCIENT AND MODERN SCORING.

CHAPTER XV.

CHAPTER XVI.

ARCHERY;

ITS

THEORY AND PRACTICE.

—o—

Chapter I.

INTRODUCTORY.

But little apology is, I think, needed, for presenting to the lovers of the long bow the present Work upon the Theory and Practice of Archery. The rapidly increasing taste for this elegant and manly amusement (requiring, as it does, both physical powers and mental study for its successful practice), and the eager desire of many to excel in this their favourite pursuit, seem to call for some more practical and scientific Treatise upon the Art, than at present exists. No disparagement is here intended to the clever and amusing Works upon the same subject already before the public; but it is an undeniable fact—and the opinion of almost every experienced Archer can be adduced to bear me out in asserting it—that all those Works, without exception, fail to touch upon or develop any fixed theoretic principle of shooting, and totally ignore those more abstruse and delicate points connected with its practice, upon which accurate and scientific Archery mainly depends. This may appear a somewhat bold and presumptuous assertion to make, but that it is a fact, few, who have endeavoured to find written instruction to guide them in the pursuit, will be tempted to deny; and the principal reason for its being so would appear to be, that, at the time these publications appeared, the knowledge of the Art and the powers of the bow had either been partially lost, or had not reached such a state of development as is the case at the present day—consequently, much less being known about it, much less could be taught. Just one example shall be mentioned in corroboration of this view of the matter. Mr. Roberts, in his very talented Treatise on Archery, published in 1801 (perhaps the best at present extant), records the following performance, as being one of what was considered in his time the great feats of the day, namely, that in one hundred arrows, shot at the distance of one hundred yards, fifty-two actually struck the target! Wonderful, indeed! Is there any third-rate Archer of the present day who has not done as much, and a great deal more, over and over again? The name of many a brother Archer occurs to me at this moment, who would be exceedingly disappointed, indeed, at having, in any morning's practice, *only* achieved such a performance as this! but

Tempora mutantur, nos et mutamur in illis.

Archery of 1856 is not the Archery of half a century, or even of twenty years ago. Scores that would then have been deemed impossible and visionary, are now of every-day occurrence; and the Robin Hoods and Little Johns of those days, could they but be pitted against the present living magnates of the bow, would occupy but a sorry position indeed.

Another reason to account for the undoubted omission in these Works of practical and scientific instruction, may be here further adduced; namely, that their authors were, with but few exceptions, themselves Archers of no note even in their own days, and therefore, not the best qualified for its exposition, even up to the standard of knowledge at that time attained.

Having thus far demonstrated that the want exists of a practical Work on Archery by a practical Archer, and being well convinced of it by my own experience, and supported in that conviction by the almost unanimous opinion of my brother Toxophilites, I am emboldened to lay before the public the following Treatise, containing the results of considerable experience and much hard study; as likely, it is hoped, in some measure to supply the deficiency complained of, and to prove of practical utility not to the youthful aspirant only, but to Archers of more advanced experience also.

In the course of these pages, whilst giving utterance to my own settled convictions, some things must of necessity be said, some positions advanced and rules advocated, that will, it is feared, jar considerably with the preconceived opinions and time-honored prejudices of many of the lovers of the bow. Even some of the revered dogmas of good old Roger Ascham, and of a period antecedent to him, hitherto received as absolute, and the mere doubting of which will appear to many the very height of presumption and little short of rank heresy, cannot be wholly subscribed to. The doctrine of the necessary superiority of old ideas over new ones, though supported by no reasoning, no argument whatever, and resting on the bare assumption only that, as our forefathers did so, therefore *ex necessitate rei* we their descendants should do so likewise, will still find advocates, even in these our times of progress and knowledge. To such, then, I would merely remark that, inasmuch as very great success has in my own case, and in that of other leading Archers of the day, followed the adoption of the theories and system hereafter laid down, so these may fairly be presumed to possess peculiar claims to careful attention, and to deserve something better than a hasty rejection on account of an apparent antagonism to preconceived notions.

It has generally been the custom to commence a Treatise on Archery with an elaborate defence of its practice; and into such contempt, until of late years, had it, indeed, fallen amongst the non-shooting members of society, as being in their opinions a mere childish amusement, that this was both called for and unavoidable on the part of any author, desirous of disabusing the public mind of a groundless and ridiculous prejudice.

At the present day, however, such a defence can hardly be considered necessary; for since the establishment of the Grand National Archery Society, some fourteen years back, the knowledge of the real powers of the bow, and the qualities required for their cultivation, having become more universally known and better appreciated, this prejudice may be said to have died away; and, in addition to this, were such a defence, indeed, required, it has been already ably and successfully supplied by several of my predecessors. A few points may, however, be additionally touched upon, and to these, previous to the conclusion of these pages, I shall more particularly address myself.

Before closing this introductory chapter, let me address a few words of advice and encouragement to the beginner. First of all, make up your mind to succeed, for that is one of the best elements of success in everything; and, secondly, expect plenty of difficulties and discouragements, for you will be sure to meet with them. It is not easy to become great in any thing, and Archery forms no exception to the rule. Do not go hunting about for a royal road to the bull's eye—none such exists—you must work hard and practise regularly, before even moderate success will reward your efforts. Use your brains as well as your muscles—study as well as practice. Brute force alone will never make an Archer. Above all, do not fancy yourself a first-rate shot, when you are only a *muff*—nothing will so much tend to keep you one all your days as this. A mistaken vanity is the very bane of all improvement. Having once passed the *pons asinorum* of Archery, you will begin at once to taste its pleasures. There is no exercise more healthy or more rational, or which returns more true and genuine gratification to the man who practises it. A well shot arrow lodged in the right place not only pleases the spectator, but is a source of unmingled gratification to the shooter also. May the study of these pages assist you in attaining, as often as possible, this most desirable end.

Chapter II.

A Glance at the Career of the English Long-Bow.

The Anglo-Norman Period—Robin Hood—Military Achievements of the Bow in the Middle Ages—Its Decline and Fall—Revival for Amusement—First Toxophilite Societies—Establishment of National Archery Society.

Although, as stated in the last chapter, this work is essentially intended to treat on the practical and scientific points connected with the pursuit of Archery, and mainly thereby to supply an acknowledged want, still it may not be altogether uninstructive to present to the reader a short and compendious sketch of the history, powers, and doings of the bow, from the time of its first introduction into this country to the present day; so that every Archer may have a general knowledge of the career of this his favourite weapon, and be able to render a reason for the high estimation in which he doubtless holds it. Those that are desirous of more detailed information are specially referred to Mr. Roberts's work already alluded to—a work abounding in extracts from every author of authority who has written on the subject, and containing a mass of information that will amply repay a careful perusal.

The date of the first introduction of the long-bow into England is a matter of considerable uncertainty, and a *cheval de bataille* with all historians and authors who have attempted to determine it; but it is certain that it was not till after the battle of Hastings, and the subsequent conquest of Britain by the Normans, that it became the favourite and specially encouraged military weapon in the hands of its inhabitants. The preponderance of historical evidence goes to prove that, to the deadly effects produced by it in that battle, the invaders principally owed their victory—Harold himself and the best of his men falling victims to the clothyard shaft. Thus the long-bow proved the prime agent in subjugating this country, substituting the Norman for the Saxon rule, and, by the intermixture of the two people, ultimately in completing that far-famed Anglo-Saxon race, the popularly supposed powers of which to accomplish everything everywhere it behoveth not one of themselves further to dilate upon. From this time, then, we may conclude, commenced in England that general, and all but universal, cultivation of the bow, which was ultimately to lead to such marvellous and astounding results, and to render the very name of the English bowman an object of terror and dread in the minds of his enemies. Archers we find employed on both sides in the civil contests between Stephen and Matilda, and during the reign of Henry II. they began to form the larger portion of the infantry of the English armies, and to evince that decided superiority over those of every other nation which they ever afterwards retained.

In this reign, too, first appeared upon the scene that prince of good fellows (as times went) and gentlest of robbers and outlaws, bold Robin Hood!—that hero of impossible shots, the twang of whose bow, with that "of his jolly companions every one," could, according to Drayton, be heard a mile off! *Credat Judæus!* However this may be, if there be truth at all in history and legend, he and his merry men were incomparable Archers, for strength and skill never surpassed, if ever equalled; and we may well suppose Archery to have been brought to the highest pitch of perfection in the times that produced such eminent exemplifiers of the Art. Robin flourished much longer than is usual with such bold spirits, even in the olden time; for we find him still in his glory through the reign of Richard I., John, and a considerable portion of that of his successor, Henry III.

It would be impossible, without entering into a mass of details whose length would be unsuited to the nature of these pages, to mention a tithe of the extraordinary feats performed and victories gained by the English during the next three or four centuries, owing entirely to their superiority in the use of the long-bow. The fictions of romance pale before many of the authenticated tales handed down to us by historians of the wonders it achieved. No armour that could be made proved strong enough to insure its wearer against its power, no superiority of numbers seemed sufficient to wrest a victory from its grasp. Speed declares that the armour worn by Earl Douglas and his men-at-arms at the battle of Homildon had been three years in making, and was of remarkable temper, yet the "English arrows rent it with little adoe." Gibbon tells us that, on one occasion, during the Crusades, "Richard, with seventeen knights and three hundred archers, sustained the charge of the whole Turkish and Saracen army;" and the pages of Froissart teem with the details of battles and skirmishes without number, in which the irresistible power placed by it in the hands of the English enabled them to set all odds at defiance, and constantly to emerge victorious out of situations where utter destruction seemed certain and inevitable. Look at Cressy and Poictiers, Navaretta and Agincourt! Since the extinction of the bow as a weapon of war, has England ever shown parallels to such victories as these? "Let him," says Roberts, "who reads the history of modern times, look narrowly to find, if but once (since Archery flourished), with our *twelve or fifteen thousand* we have defeated an army of *fifty or sixty thousand;*" and he might have added, as was the case in the last-named battle, if with *twenty-five thousand* we had completely routed and nearly annihilated an army of a *hundred and sixty thousand!* And be it also borne in mind that these marvellous and wondrous results were not obtained against barbarian hordes or undisciplined soldiery, but against some of the first chivalry and most renowned men-at-arms that the world at that time contained. In spite of Miniés and breech-loading rifles, will it ever again become a proverb in vogue regarding the British soldier, that he carries as many enemies' lives in his hands as bullets in his pouch; yet it was a common saying in Scotland in times gone by, that every English Archer bore with him the lives of four-and-twenty Scots—such being the number of arrows each carried in his quiver. All honour, then, to the long-bow! May the grateful remembrance of it never pass away from the land, whose glory it has raised to so high a pitch; and though it may never be seen a weapon of war again, may its practice long continue to form one of our most manly and health-inspiring amusements.

The time that Archery commenced its decline in this country, till it finally ceased to be used in warfare at all, is almost as much a matter of dispute with writers as is the date of its first introduction. If we are to believe Moseley, "the battle of Agincourt (which happened under Henry V., 1415) is the last important action in which Archery is mentioned;" but according to Roberts, (whose accuracy in matters of historical detail can in general be well depended on,) great slaughter was caused by it in the civil wars between the White and Red Roses; and he further adds, "it continued to support its military character and invincible career of glory with undiminished effect during the reigns of Henry VI., Henry VII., Henry VIII., and Edward VI., and even in the reign of Elizabeth was still in high repute amongst foreigners of great military skill, who had witnessed its powerful effects." Nevertheless, we find Hollingshead, who wrote in the sixteenth century, bewailing the degeneracy of the Archery of his day, as being deficient in force and strength. The mean between the extremes of conflicting opinions will probably lead us to the nearest approximation to the truth. It may, therefore, be concluded that towards the close of the fifteenth century the use of fire-arms had caused Archery to be held in somewhat less repute than formerly, and that, consequently, the cultivation of it had ceased to be of that all but universal character that it once had been. The natural effects followed—with less practice came less strength and skill; and by the time the sixteenth century came to an end, but little remained to the bow, beyond the remembrance of its former glory and achievements. The last mention of Archery as used in warfare, occurs in a pamphlet published in 1664, where it is stated to have been employed in the contests between the Marquis of Montrose and the Scots; but evidently for many years prior to this date, its ancient pith, power, and reputation, had departed.

We now arrive at the time when the bow, abandoned as a weapon of war, became a mere instrument of amusement and recreation; but hardly any record exists to enlighten us as to the extent to which it was practised, or the degree of skill retained by its admirers. During the eighteenth century it would almost appear to have fallen entirely into disuse, only two or three societies existing in the kingdom, and those in a very languid and feeble condition. In the year 1780, however, a society, under the title of The Royal Toxophilites, was established in London; and, the impetus once communicated, a great revival of Archery immediately took place, and a vast number of societies speedily sprang up in every part of the country, the greater part of which, with many new and more modern ones, exist in full force and vigour at the present day. Undoubtedly, however, we owe to the establishment of the Grand National Archery Society, fourteen years back, the present high consideration in which the practice of Archery is by both sexes now held, as well as the more general and increasing skill which continues year by year plainly to manifest itself—thus showing that the love of the bow has only slumbered, not died, in the breasts of Englishmen, and needs but moderate encouragement to become once more, if not a weapon of war, at any rate one of the most esteemed and highly-prized amusements in the kingdom. To conclude, let every Briton remember, in the words of Camden, that when Englishmen used Hercules' weapons—the *bow* and the black bill—they fought victoriously, "with *Hercules' success*"—and reverence their memory accordingly.

Chapter III.

Of the Bow.

Varieties of Form and Material—The Flodden Bow—The Bows in the Tower—The Self-Bow, its Form, Texture, and Weight—Quality of the Yew—The Backed-Bow—Woods mostly employed—The Shape—Cause of the Jar—The Length—Relative Merits of the Self and Backed Yew Bows—The Carriage Bow.

Of the various implements of Archery the bow demands the first consideration, and to it I shall therefore devote the present chapter. A general, though necessarily brief, outline of its reign and use in this country, and of its power and character in the hands of the English, having already been given, it may only be necessary further to add that, in almost every nation, it has, at one period or other, formed one of the chief weapons of war and the chase, and is, indeed, at the present day, in use for both these purposes in various parts of the world. It has differed as much in form as in material, having been made curved, angular, and straight; of wood, metal, horn, cane, whalebone only, or of wood and horn, or wood and the entrails and sinews of animals and fish combined; sometimes of the rudest workmanship, sometimes finished with the highest perfection of art. But, as it is certain that in no country has the practice of Archery been carried to such a degree of perfection as in our own, so is it equally undeniable that no bow of any other nation has ever surpassed or, indeed, equalled the English long-bow in respect of strength, cast, or any other requirement of a perfect weapon. This being an indisputable fact, it would be a waste of space and a departure from my immediate object were I to enter into a description of the bows used at various times in different countries, or into a discussion as to their respective merits. I shall not, therefore, do so, but confine myself to the practical point of treating upon the English long-bow, that being still, as it always has been, the only one in use and favour in this country. The cross-bow is, of course, altogether a different instrument. It is a matter for surprise and regret that so few, if any, genuine specimens of the *old* English long-bow should remain in existence at the present day. The only one with which I am acquainted was, and, I believe, still is, in the possession of Mr. Muir, of Edinburgh, said to have been used at the battle of Flodden, in 1513, is of self-yew, apparently of English growth, and very roughly made. Its strength is *supposed* to be between 80 and 90 lbs.; but, as it cannot be proved without great risk of breaking (a risk its owner is very properly unwilling to run), this is matter of supposition only. This bow was presented to Mr. Muir by Colonel J. Ferguson, who obtained it from a border house, contiguous to Flodden Field, where it had remained for generations with the reputation of having been used at that battle. The specimen is probably unique, and has every appearance of being genuine.

The Flodden Bow.

There are likewise in the Tower at the present time two bows taken out of the "Mary Rose," a vessel sunk in the reign of Henry the Eighth. They are rough, unfinished weapons, quite round from end to end, tapered from the middle to each end, and without horns. It is difficult to estimate their strength, but they do not *appear* to exceed 65 or 70 lbs.

Before proceeding to the discussion of the practical points connected with the bow, I must beg my brother Archers to bear in mind, once for all, that these pages profess to give the result of actual experience; and to assure them that nothing to be advanced in them is mere theory, or opinion unsupported by proof, but is the result of long, patient, and practical investigation, and of constant and untiring experiment. Whenever, therefore, one kind of wood, or one shape of bow, or one mode or principle of shooting, &c., &c., is spoken of as being better than another, or the best of all, it is asserted so to be, simply because, after a full and fair trial of every other, the result of such investigation bears out that assertion. No doubt but some points contended for will be in opposition to preconceived opinions and practice, and will be set down as innovations—and so perhaps they are. The value of theory, however, is just in proportion as it can be borne out by practical results; and, in appealing to the success of my practice as a proof of the correctness of the opinions and principles upon which it is based, I am moved by no feeling of conceit or vanity, but wholly and solely from the desire of giving as much force as possible to the recommendations put forth, and to extort, even from my opponents themselves, at least a fair and impartial trial of them, previous to their being condemned. With these preliminary observations, (which will apply generally to the whole course of this work,) I will proceed with my subject.

The English bows in use at the present day may be divided into two classes—the self-bow and the backed-bow; and to save space and confusion, I shall confine my remarks at present to the former, reserving for hereafter anything to be said respecting the latter;

premising, however, that much to be said of the one applies equally to the other: the discrimination of my readers will at once distinguish where this occurs.

The self-bow is the real old English weapon; the one with which the many mighty deeds that rendered this country renowned in times gone by were performed; for, until the decline and extinction of Archery in war by improved fire-arms, and the consequent cessation of the importation of yew staves, backed-bows were unknown. Ascham, who wrote in the sixteenth century, when Archery had degenerated into little else than an amusement, mentions none other than selfs, and it may be, therefore, concluded that such only existed in his day. Of the woods for self-bows, Yew beyond all question carries off the palm; other woods have been, and still are, in use, such as Lance, Cocus, Washaba, Rose, Snake, and some others; but they may be summarily dismissed with the remark, that self-bows made of these woods are all, without exception, radically bad, being heavy in hand, apt to jar, comparatively dull in cast, and very liable to chrysal and break, and that no Archer should use them so long as a self-yew or a good backed-bow is within his reach.

The only wood, then, for self-bows, I may say, is Yew, and the best Yew is of foreign growth, though occasionally staves of English wood are met with which almost rival it. This, however, is the exception; as a rule the foreign is best; it is cleaner and finer in the grain, stiffer and denser in quality, and requires less bulk in proportion to the strength of the bow.

The great bane of Yew is its liability to knots and pins; and rare, indeed, it is to find a six foot stave without one or more of these undesirable companions. Where, however, a pin does occur, it may generally be rendered harmless by the simple plan of

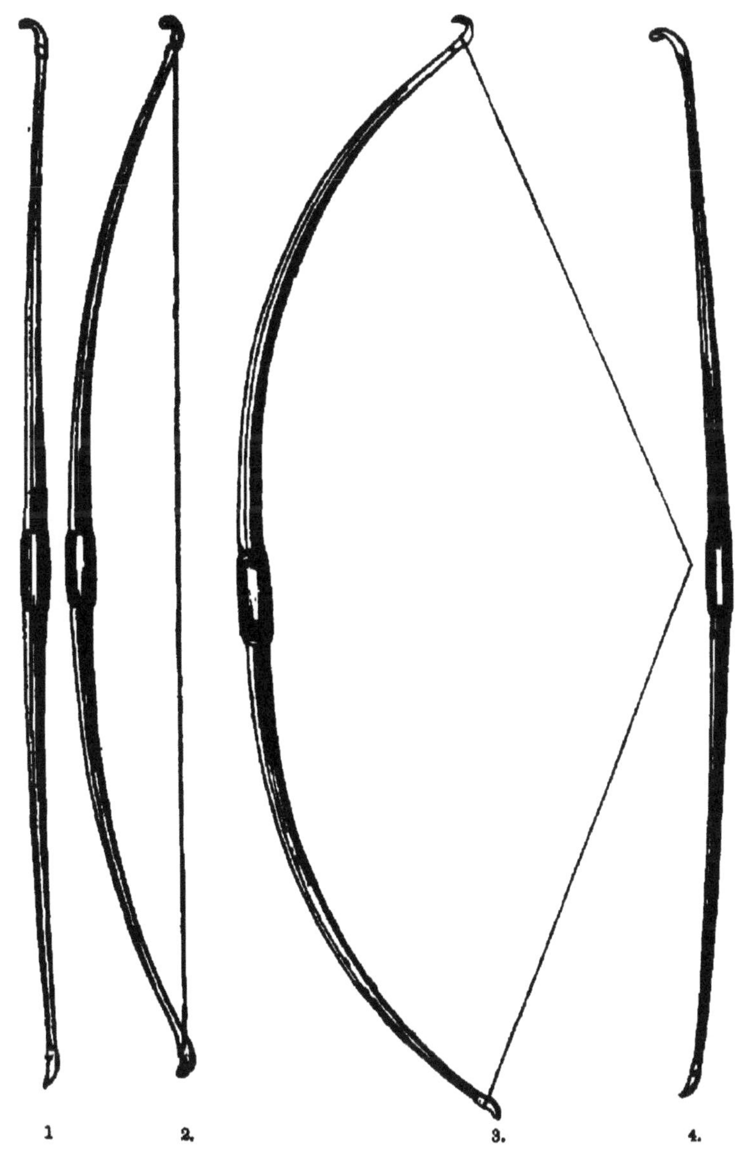

From photographies of actual bows.

1 An excellent shape.
2. Ditto ditto when strung 6 inches.
3. The correct bend when drawn 27 inches
4. A reflexed bow, and one that bends in the hand (bad).

PLATE III.

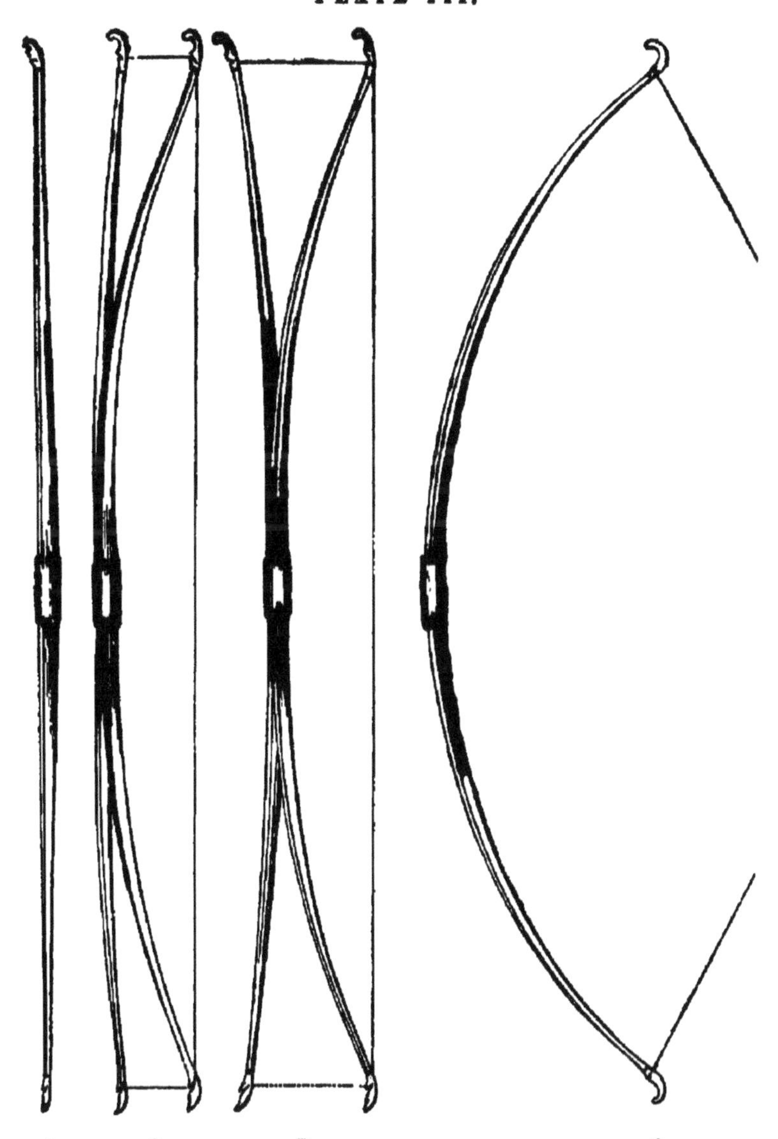

5. A good shape for a new bow After use this will come to follow the string a little.

6. and 7. Show the different distances which the limbs of a reflexed bow and of a bow that follows the string have to go to their rest.

8. A reflexed bow that bends from end to end drawn 27 inches. A very common shape, but the very worst, as it will jar and kick in the hand and

"raising" it, *i.e.*, leaving a little more wood than elsewhere round the pin in the belly of the bow. This strengthens it, and diminishes the danger of a chrysal (which is a small crack attacking the weak places, almost imperceptible at the commencement, but which, by degrees, enlarges itself, and ultimately eats into the bow, as it were, until it breaks). The grain of the wood should be as even and fine as possible; not cross, nor running out towards the middle, nor winding. It should be thoroughly well seasoned, and of a good, sound, hard quality. The finest grain is, undoubtedly, the most beautiful and uncommon; but the open or less close, if straight and free from knots and pins, is nearly, if not quite, as good for use.

The self-bow may be made of one single piece, or of two pieces dovetailed together in the handle. If of one piece, the quality of the wood will not be quite the same at both ends, the lower part being slightly denser than the upper; whilst the grafted bow may be made of the same piece, cut or split apart, and so of exactly the same nature. The difference, however, is so slight as to be immaterial. Care must be taken, in choosing a grafted bow, to see that it be put firmly together in the middle.

In shape, the bow should be full in the centre, and taper gradually to each horn; not bend in the hand, or the cast will be deficient, and it will most likely jar in addition. (See plates No. 2 and 3.) A perfectly graduated bend from a stiff centre to each horn is best. Some self-yew bows are naturally reflexed, others quite straight, and others, again, follow the string. The reflexed are more pleasing to the eye, but liable to the above objection of jarring. Those which follow the string a little are the most pleasant to use.

The handle, which should be regulated to the grasp of each Archer, ought to be in such a position that the upper part of it may be from an inch to an inch and a quarter above the true centre of the bow: if placed in the exact middle, the bow will be apt to kick. If it be grasped properly (inattention to which will often cause the lower limb of the bow to be pulled out of shape), the fulcrum in drawing will be about the centre. The upper limb, being thus a little longer than the lower, must of necessity bend a trifle more, and this it should do. For covering the handle, nothing is better than green plush.

It is customary to let into the bow, just above the upper part of the handle where the arrow lies, a small piece of mother-of-pearl, ivory, or other hard substance. This serves to prevent the wearing away of the wood by the friction of the arrow, which is greater or less according to the slope of the bow, and the attention or otherwise of the Archer in wiping his arrows when needed.

The length of the bow is here calculated from nock to nock, and should be regulated by its strength, and the length of the arrow to be used with it. As a rule for safety, I should say the stronger the bow the greater should be its length; and so also the longer the arrow, the longer the bow. For those who draw the usually 28-inch arrow in bows of from 48lbs. to 55lbs., a useful and safe length would be about 5ft. 10½ in. If this length of arrow or weight of bow be increased or decreased, then let the length of the bow be proportionably increased or decreased also, taking as the two extremes, 5ft. 7in. for the shortest and 6 feet for the longest. I would have no bow outside of either of these measurements. It may be here remarked that a short bow will, perhaps, cast further than a longer one of the same weight; but this extra cast is only gained at a greater risk of breakage. As bows are generally weighted and marked for a 28-inch arrow, a greater or less pull than this will take more or less out off them; and the Archer's calculations must be made accordingly.

To increase or diminish the power of a bow, the usual plan is to shorten in the one case, and reduce (in bulk) in the other. In all cases the horns should be well and truly set on, and the nocks be full and round. If the edges be sharp, the string will, in all probability, be cut, and, in consequence, break sooner or later, and endanger the safety of the bow.

I now come to the second part of my subject, namely, the backed bow. From all that can be learnt respecting it, it would appear that its use was not adopted in this country until Archery was in its last state of decline as a weapon of war, when, the bow degenerating into a mere instrument of amusement, the laws relating to the importation of yew staves from foreign countries were evaded, and the supply consequently ceased. It was then that the bowyers hit upon the plan of uniting a tough to an elastic wood, and so managed to make a very efficient weapon out of very inferior materials. This cannot fairly be called an invention of the English bowyers, but an adaptation of the plan which had long been in use amongst the Turks, Persians, Tartars, Chinese, and many other nations, more especially the Laplanders, whose bows were made of two pieces of wood united with isinglass. As far as regards the English backed-bow (this child of necessity), the end of the sixteenth century is given as the date of its introduction, and the Kensals, of Manchester, are named as the first makers—bows of whose make are still in existence and use, and are generally made of Yew, backed with Hickory or Wych-Elm.

The backed-bows of the present day are made of two or more strips of the same or different woods glued and compressed together, as firmly as possible, in a frame with powerful screws, which frame is capable of being set to any shape. Various woods are used, all of which make serviceable bows, though differing much in quality, For the back we have Lance, Hickory, American and Wych Elm, Hornbeam, and the sap or white part of the Yew; for the belly, Yew,

Washaba, Lance, Snake, Fustic, and some others inferior to these, are used. But of all combinations it may be said, "*Micat inter omnes* Yew-backed Yew, *velut inter ignes Luna minores.*" This is the real rival of the self-yew, the one that stands pre-eminently forward in the ranks of the backed, the disputer of its supremacy; but more of this by-and-bye, when comparing the respective merits of the two bows. Then next in quality comes Yew backed with Hickory, or any other tough wood; and then, *longo intervallo,* Fustic, Washaba, and Lance, backed in like manner. For bows of three pieces, Yew, Fustic, and Hickory, will hardly be improved by any other combination; but, as a general rule, bows of two pieces are preferable, as the more glue there is about a bow the more the danger exists of a breakage from damp, and in no one point does a bow of three or more pieces excel one of two.

The next point to be treated of is a most important one, namely, *the shape;* and here I shall differ most materially from the commonly received opinion. The backed-bow is generally made reflexed, and bends in the hand, more or less, according to the amount of the reflex. (See Nos. 4 and 8—Plates 2 and 3.) Now the exact reverse of this is contended for, and it is boldly maintained that every particle of reflex is bad, and that the proper shape is either *straight* or a trifle *following the string*—similar, in fact, to that before recommended for the self-bow, namely, full and stiff in the centre, and tapering gradually to each horn. The first quality of a bow is *steadiness;* now every degree of reflex is accompanied by a like degree of jar or kick, the effect of which causes the very reverse of this quality; and this holds good equally in respect of self-bows, which are sometimes, though rarely, naturally reflexed, and sometimes purposely so set when grafted, though the naturally reflexed self yew-bows do not generally retain that shape for any length of time, but, with a little use, come to the string so far as to do away with the unpleasant jar.

The jar or kick in reflexed bows has always appeared to me to arise from the following cause: when the bow is set free by the loose, its natural elasticity causes it to return as far as it can to its original shape, so that the further each limb has to go to its rest the greater becomes the struggle when checked by the string. (See Nos. 6 and 7. Plate 3.) This is shown by the fact that reflexed bows are almost invariably broken by the fracture of the string, whilst the contrary is the case with those which follow it. The less then there is of that violent struggle (so to speak) on the recoil, the less there will be of the jar or kick, and the steadier in consequence the shot. This may be easily tested by shooting a few dozen arrows with a bow that follows the string, and immediately afterwards with a reflexed one. A man must be prejudiced indeed who will not allow that there is a vast difference between the two upon the point in question. Now what can be urged in favour of the reflex? Has it any peculiar merit of its own to compensate for the absence of this first element of a good bow—steadiness? Even its strongest advocates can only assert in its favour that it adds to the spring; but granting that this is so (which I do not), are a few extra yards of cast worth gaining at the expense of the finest quality a bow can possess, and without which accurate shooting is impossible? The reflex, too, adds materially to the chance of breaking both by chrysals, damp, and the fracture of the string, as the wood, particularly of the belly, is forced out of its natural shape, recoils farther, and meets with a more violent check when stopped by the string; so that, even supposing it gains a trifle on the point of cast, it loses infinitely more on the two equally important ones of steadiness and safety. I think that no one will be tempted to deny that the best form of bow is that which is steadiest in cast, freest from jar or kick, and pleasantest and safest in use; and that, it is confidently affirmed, *is not* the *reflexed.*

Now comes a question which may well admit of dispute, and

which must, after all, be left to each Archer to decide for himself. Which is best: a well-made self-yew, or an equally well-made yew-backed yew-bow? (Other backed-bows, though good and serviceable, especially yew-backed with hickory, I cannot think come up to these.) The advocates of the self-yew affirm their pet weapon to be the sweetest in use, the steadiest in hand, the most certain in cast, and the most beautiful to the eye; and in all these points, with the exception of that of certainty of cast (in which respect the yew-backed yew is fully equal) they are borne out by the fact. This being the case, how is it then that a doubt can still remain as to which is most profitable for an Archer to make use of? Here are three out of four points (two of which are most important) upon which it is admitted the self-yew is superior; and yet, after much practical and experimental testing of the two bows, I hesitate to which to give the preference, and knowing not which to recommend, must, after all, as before said, leave it to the taste and judgment of every man to decide for himself. The fact undoubtedly is, that the self-yew is the most perfect weapon; but it is equally an undoubted fact, that it requires more delicate handling than its rival: since, its cast lying very much in the last three or four inches of the pull, any variation in this respect, or difference in quickness or otherwise of the loose, *varies the elevation of the arrow to a much greater extent* than the same variation of pull or loose in the backed-yew, whose cast is more uniform throughout. Now, were a man perfect in his physical powers, or always in first-rate shooting condition, there would be no doubt as to which bow he should use, as he would in this case be able to attain to the difficult nicety required in the management of the self-yew; but as this never can and never will be, the superior merits of this bow are partially counteracted by the extreme difficulty of doing justice to them; and, the degree of harshness of pull and unsteadiness in hand of the yew-backed yew being but trifling, the greater certainty with which it accomplishes the elevation counterbalances, upon average results, its inferiority in other respects. Another advantage the self possesses is, that it is not liable to injury from damp, when the backed is; but then the latter costs little more than a third of the money, and with common care need fear no harm from that cause; an inch or two of lapping at either end, close to the horns, will go a long way to preserve it from this danger. As regards chrysals and breakage from other causes than damp, neither possesses any advantage over the other. The main results of the different qualities of the two bows resolve themselves into these two prominent features, namely, that the self-yew bow, from its steadiness, sweetness, and absence of vibration, ensures the straightness of the shot better than the backed-yew; whilst the latter, owing to its regularity of cast not being confined to a hair's-breadth of pull, as it were, carries off the palm for certainty of elevation, and this favourable attribute belongs to backed-bows generally.

As regards backed-bows other than yew, it has already been observed that they are inferior to the two sorts just treated of. But it must not be supposed from this that it is intended to affirm that they are bad or unfit for the Archer's use—on the contrary, if properly made, they are good and serviceable weapons, only less to be recommended than the two kinds of yew-bows; neither must the idea be adopted from what has been said respecting the superiority of yew as a wood for bows, that therefore *all* yew-bows are necessarily good or better than those of other woods; such is far from being the case, for a backed-bow, *well made* of a good piece of Fustic, Washaba, or Lance, is decidedly better than either a self or backed one made of inferior yew. It is only to the best samples of yew-bows of either kind that the foregoing remarks are intended to apply.

There is a bow called the "carriage-bow," which here requires some notice. It is made to divide in the centre by means of an iron or brass socket fixed to the lower limb of the bow—something similar to the joint of a fishing-rod, in fact. The only object attained, however, is that it enables the Archer when travelling to carry his bow in a smaller compass: but to obtain this, much additional weight is added to the bow, rendering it heavy in hand, and unpleasant in use. The remedy here, therefore, is worse than the disease.

The Carriage Bow.—A. The Socket.

Chapter IV.

How to Choose a Bow, and How to Use and Preserve it when Chosen.

Popular Errors in the Choice—Most Accomplished Shots—Causes of Success and Failure—Principles Guiding the Selection of a Bow—Its Preservation and Repair—Unstringing.

The next point to be considered with reference to the bow is the strength to be chosen; and, respecting this, the first thing to be observed is that it must be completely under the shooter's command —within it, but not much below it. One of the greatest mistakes young Archers commit (and many old ones too) is that they *will* use bows too strong for them. (How many of us, by-the-bye, are there to whom, at one period or other of our Archery existence, this remark has not applied?) The natural desire to be considered strong and muscular appears to be one of the moving agents to this curious hallucination, as if a man did not expose his weakness more by straining at a bow evidently beyond his strength (and thereby calling attention to his weakness) than by using a lighter one with grace and ease, which always gives the idea of force, vigour, and power. Another incentive to strong bows is the passion for "sending down the arrows sharp and low," and the consequent using of powerful bows to accomplish it; the which is, perhaps, a greater mistake than the other, for it is not so much the strength of the bow as the perfect command of it that enables the Archer to obtain this desideratum. The question is not so much as to what a man can *pull* as to what he can *loose;* and he will without doubt obtain a lower flight of arrow by a lighter power of bow, under his command, than he will by a stronger one beyond his proper management.

How many a promising Archer has this mania for strong bows destroyed (in an Archery sense of the term). I call to mind one at this moment—one of the best and most beautiful shots of his day; a winner, too, of the second and first prizes at the Grand National Meeting two successive years—whose accuracy was at one time completely leaving him, and dwindling beneath mediocrity, owing, as I firmly believe, to his infatuation upon this point. Another I had a slight acquaintance with brought himself to death's door, by a violent illness of nearly a year's duration, by injury to his physical powers, brought on by the same thing, only carried to a much greater excess. And, after all, the thing desired is not always obtained.

Let me transport my reader, in mind, to any field where the Annual Grand National Archery Meeting is held. Observe, there are from eighty to a hundred picked shots of the country standing at the targets, contending with all their might for the prize of honour and skill. Whose arrows, think you, fly down the sharpest, the steadiest, the keenest? Are they those of the sixty and seventy pounders? Not a bit of it: observe that Archer from an eastern county just stepping so unpretendingly forward to deliver his shafts. See with what grace and ease the whole thing is done—no straining and "contortioning" here. Mark the flight of his arrow! how keen and low, and to the mark! None fly sharper, few so sharp; and what, think you, is the strength of that beautiful self-yew he holds in his hand? Why, 50lbs. only! and yet the pace of his shaft is unsurpassed by any, and it is nigh upon five shillings in weight too. Here is another; mark his strength and muscular power—60, 70, or even 80lbs. are probably within his pull, yet he knows better than to use such bows where the prizes are awarded for skill, not brute force. The one he shoots with is but 48lbs., yet how steady and true flies the arrow! how charming in its flight! And so on, all through the field, you will find it is not the strong bows, but those that are under the perfect command of their owners, that do their work the best.

Inasmuch, then, as the proper flight of the arrow from any bow depends almost entirely upon the way in which it is loosed, the strength of the bow must not be regulated by the mere muscular powers of the individual Archer: for he may be able to draw a 29-inch arrow to the head in a bow of 70 or 80lbs. without being able, after all, to loose steadily during a match more than 56 or 60lbs., if as much. Not the power of drawing, but of loosing steadily, must, therefore, be the guide here. Let the bow be within this power, but well up to it; for it is almost as bad to be *under* as *over* bowed. The evils attendant on being over-bowed are various; the left arm, the fingers of the right hand, and the wrist are strained and rendered unsteady; the pull becomes uncertain and wavering, and never twice alike; and the whole system is over-worked and wearied, and the mind depressed by ill success—the entire result being disappointment and failure. On the other hand, care must be taken not to fall into the opposite extreme of being *under-bowed*, as in this case, also, the loose becomes difficult, and generally unsteady and unequal. The weight of bows now in use varies generally from 48 to 56lbs., the weaker or stronger ones forming the exception, and this weight is ample for the distances usually shot, which very rarely exceed 100 yards. Let each, therefore, find out what he can draw with ease and loose with steadiness during a day's shooting, and choose accordingly. If a beginner, probably 50lbs. is the outside weight he should commence with; a pound or two less in most cases would, in all likelihood, be even better. This is, however, a matter that alone the individual Archer can determine for himself.

It is best always to use the same weight of bow and length of arrow; and, therefore, every Archer, if he shoot much, should possess two or three bows as much alike as possible, and use them alternately. This will prove economy in the end, as each will have time to recover its elasticity, and will last a much longer time.

To choose a bow, let each go to the maker he likes best, name the price he can afford, and the sort and weight he prefers. He will then see what choice he has. If there appears to be one likely to suit him, let him (after examining the wood, and seeing that it is free from flaws, string it,) and, placing the lower end on the ground in such a position that the string shall be under his eye and uppermost, notice whether it be perfectly straight; if so, the string, when brought to bear on the middle of the handle, will divide the bow from horn to horn into two equal parts. Should there appear to be more on one side than the other in either limb, the bow is not straight, and should be rejected. As a general rule, the lightest wood is the quickest, the heavy the most lasting—but not always. The next step is to have the bow pulled up, so as to see if it bends evenly, and gives no sign of weakness in any particular portion. The upper limb, as before stated, being the longest, should bend a trifle the most. If there be no ready-made bow to suit, the purchaser may select a stave, and have it made to his own pattern; but, on the whole, the first plan is the best, as no one can tell how a stave will make up.

Bows are broken from several causes—by neglected chrysals, or damages to the wood, by a jerking and uneven style of drawing, by dwelling too long at the point of the arrow after it is pulled up, by the breaking of the string, by damp, and oftentimes by thoughtlessness or carelessness. Whenever a chrysal appears, it should be watched, and, if found to increase, should then be firmly lapped with hemp or string, well glued; when dry, this should be rubbed smooth, painted with oil colour to the taste of the owner, and varnished. This will keep the glue dry, and look less unsightly. Care should be taken not to make holes in the wood with the point of the arrow, nor scratch it with the buckles of bracers, or buttons of gloves, or any of the ornaments with which the Archer may adorn his person. The less of these hard substances about the shooter the better.

Breakages from a bad style of drawing, and dwelling too long on the aim, can only be avoided by adopting a better and more rational method; those caused by the fracture of the string only by being careful never to use one that is unsafe, or too much worn. A good deal depends, in this latter case, upon the moment when the string breaks. If it goes when the arrow is fully drawn, there is little hope for the bow, as there is nothing to check or break the recoil; but if it breaks when the recoil has taken place, which is generally the case, a self or backed bow that follows the string will usually escape without damage. Breakage from damp applies to backed bows only, and great indeed is the mortality amongst them from this cause. Commonly it is the lower limb that goes, as that is most exposed to damp, arising from the ground when shooting, or the floor when put away. If the weather be moist when the bow is used, let the shooter continually rub it, and when put away especially do so, with a piece of waxed cloth or flannel. A waterproof case, and an Ascham, with the bottom raised a few inches from the floor, in a dry room, are the best preservatives I know of. It is a good thing, also, to lap the bow for about an inch close to each horn, as when this is done, though the glue come undone, the wood will often escape damage, and can be made all right again by being re-glued. Lastly, carelessness and thoughtlessness break many bows, and particularly that most silly of silly habits of bending the bow the wrong way, when unstrung, in order to "get it back to its proper shape!" If it be broken from the latter cause, any jury of sensible Archers would infallibly return the celebrated verdict of "serve him right."

A yew bow that is so much damaged by chrysals, or by accident to the wood, as to be beyond being made safe by lapping, may often be mended by adding a belly or back, as the case may require. A weak bow may be strengthened in the same way, or, if either limb be broken or irretrievably damaged, and the remaining one be sound and worth the expense, another limb may be grafted on to the old one. If possible, let this be an old limb also, as the combination of new and old wood is not always satisfactory; the former, being more yielding, is apt, after a little use, to lose its relative strength, and so to spoil the proper bend of the bow. A bow that is weak in the centre, and not sufficiently strong to allow of the ends being reduced, may be brought to the required shape and strengthened by the addition of a short belly.

With regard to unstringing the bow during shooting—say a National round of 144 arrows at the three distances—a good bow will not need it if the shooting be moderately quick, excepting at the end of each of the three distances. If there be many shooters, or very slow ones, then it may be unstrung after every three or four double ends; but I am decidedly averse to unstringing after every three shots, as many do, as the constant jerk back of the wood upon its own grain must throw an increased strain upon it, besides unnecessarily taxing the muscles of the shooter.

All that has been said respecting men's bows, with the exception of the strength and length, applies equally to those used by ladies. The ordinary strength of the latter is from 24lbs. up to 30lbs., a medium between the two being about the average weight. The length of the bow is usually about 5ft. 1in. between the nocks.

It is too common a practice amongst Archers of all sorts to throw the consequences of their own faults upon the bow-makers, accusing the weapon instead of their own carelessness or want of skill; but, before this can be justly done, let each be quite certain that he has chosen his bow with care, used it with care, and kept it with care; if otherwise, any accidents occurring are ten to one more likely to be the result of his own fault than that of the bow-maker.

Chapter V.

Of the Arrow.

How to Test its Strength and Straightness—Best Materials for its Manufacture—Apparent Antagonism between the Theory and Practice of Archery, as regards its Flight, explained—Its Different Shapes—The Feathering—The Point—Varities of—Length and Weight.

The arrow is, perhaps, the most important of all the implements of the Archer, and requires the greatest nicety of make, and excellence of materials; for though he may get on without absolute failure with an inferior bow and other tackle, unless the arrow be of the best, Robin Hood himself would have aimed in vain. Two things are essential to a good arrow, namely, perfect straightness, and a stiffness or rigidity sufficient to stand in the bow, *i.e.*, to receive the whole force of the bow, without flirting or gadding, for a weak or supple, is even worse than a crooked arrow, and it need hardly be said how little conducive to shooting straight is the latter. The straightness of the arrow may be easily tested by the following simple process:—place the nails of the thumb and middle finger of the left hand so as just to touch, and with the same fingers of the right hand spin the arrow upon them; if it revolve true and steady, and close to the nail, it is straight, but if it jumps in the very least, the contrary is the case. To test its strength or stiffness, place the pile on any solid substance, holding it by the nock, and with the other hand press it gently downward in the middle. A very little experience will suffice to tell whether it be sufficiently stiff or not. An arrow that is weaker on one side than the other should also be rejected.

Arrows are either *selfs* or *footed;* the former are made of a single piece of wood; the latter, and the more preferable, have a different and harder wood dovetailed on to them at the pile end. "A shaft," says old Roger Ascham, "hath three principal parts, the stele, the feather, and the head, of which each must be severally spoken of." The stele, that is, the wooden body of the arrow, used to be, and sometimes now is, made of different woods; but for target, or indeed any other modern shooting, all may be discarded save one —*red deal*, which, when of clean, straight grain, and well seasoned, whether for selfs or footed shafts, is incomparably superior to all others. For footing, any hard wood will do; and if this be solid for one inch below the pile, it will be amply sufficient. Lance and Washaba are perhaps the best woods for this purpose; the latter is the toughest, but the former, I think, the preferable, the darkness of the Washaba having a tendency to attract the eye. This footing has three recommendations;—the first, that it causes the arrow to fly steadier, and get through a wind better; the second, that being of a harder nature than deal, it is not so easily worn away by the friction it unavoidably meets with on entering the target or the ground; and the third, that the same hardness saves the point from being broken off, should it happen to strike against any hard substance, such as a stone in the ground for instance. Before shooting is commenced, and after it is finished, let the arrows be rubbed with a piece of oiled flannel; this will prevent the paint of the target adhering to them (which otherwise it will assuredly more or less do), and save the application of sand-paper to clean them, which is objectionable on account of its wearing away the wood.

Before entering upon the question as to the best shape of the "stele" for practical use, it is necessary to say a few words concerning a point where the theory and practice of Archery, *apparently* clash; as follows:—

If a straight arrow be placed on the bowstring, the bow drawn,

PLATE IV.

Scale ½-inch to inch.

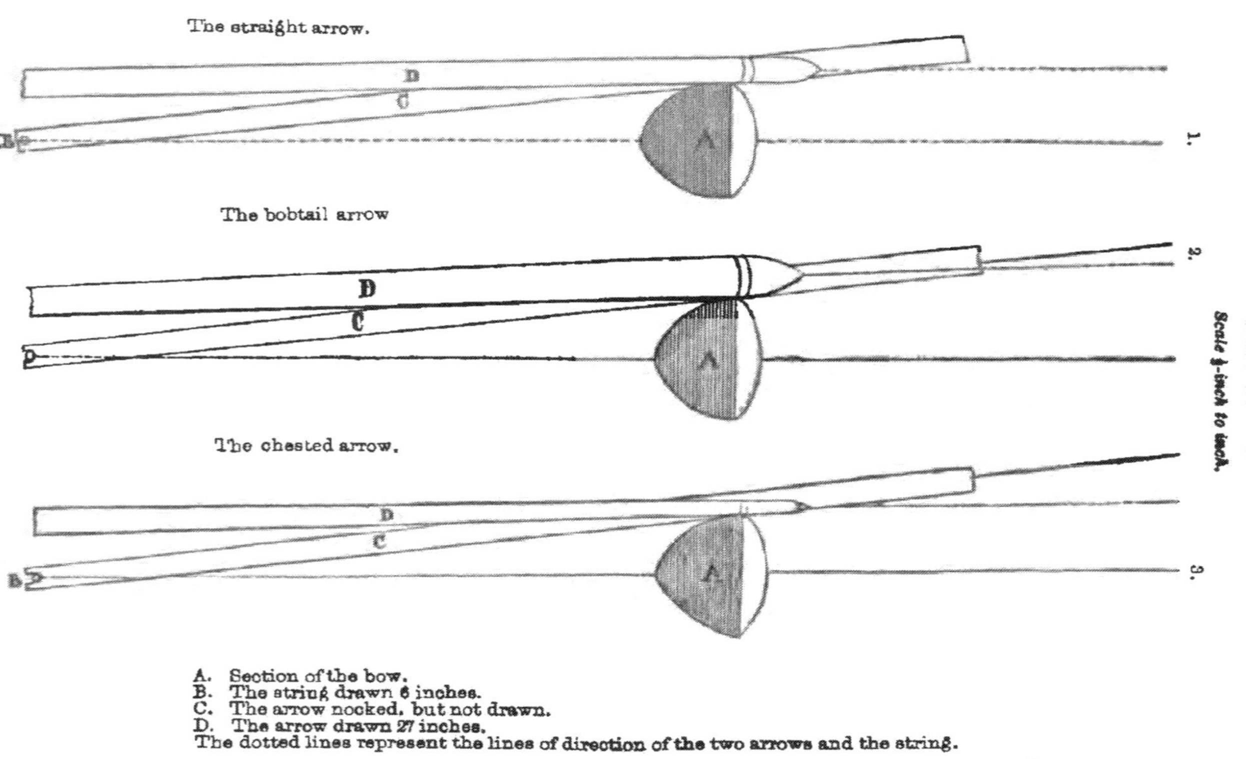

A. Section of the bow.
B. The string drawn 6 inches.
C. The arrow nocked, but not drawn.
D. The arrow drawn 27 inches.
The dotted lines represent the lines of direction of the two arrows and the string.

and aim taken at an object, and if the bow be then slowly relaxed, the arrow being held until it returns to the position of rest—*that is, if the passage of the arrow over the bow be slow and gradual*—it will be found that the arrow does not finally point to the object aimed at, but in a direction deviating to the left of it; in fact, that its direction has been altering at each point of its return to the position of rest. This is evidently due to the half-breadth of the bow, and the nock of the arrow being carried on the string, in a plane containing the string and the axis of the bow's length—and this deviation will be greater if the arrow be chested, less if it be bobtailed. (Vide plate 4).

If the same arrow, when drawn to its head, be loosed at the object aimed at—that is, *if the passage of the arrow over the bow be impulsive and instantaneous*—it will go straight to the object aimed at (the shooting being in all respects perfect).

How then is the difference of final direction in the two cases to be explained?

It must be observed that the nock of the arrow being constrained to move as it does move, causes, in the last case, a blow of the arrow upon the bow (owing to its slanting position on the bow, and its simultaneous rapidity of passage) and, therefore, a blow of the bow upon the arrow. This makes the bow have quite a different effect upon the deviation from what it had in the first case, when the arrow was merely moved slowly and gradually along it, the obstacle presented by the half breadth of the bow then causing a deviation *wholly* to the left. The *blow*, however, now considered, has a tendency to cause deviation to the left only during the first half of the arrow's passage along the bow, whilst, during the second half, it causes a deviation to the right; or, more correctly speaking, the blow of the bow upon the arrow has a tendency to cause a deviation to the left, *so long as the centre of gravity of the arrow is within the bow*, and *vice versâ*. So that, if this were the only force upon the arrow, the centre of gravity should lie midway in that part of the arrow which is in contact with the bow during the recoil.

The blow of the bow during the latter part of the arrow's passage causing deviation of the point towards the right, is, however, counteracted to some extent, if not altogether, by the action of the string which holds the arrow.

The struggle between these two forces is clearly indicated by the appearance of the arrow near the place where it is in contact with the bow when it leaves the string. It is here that the arrow is always most worn.

The nature then of the dynamical action may be thus briefly explained. The *first* impulse given to the arrow, being instantaneous and very great in proportion to any other forces which act upon it, impresses a high initial velocity in the direction of aim, and this direction the arrow recovers, notwithstanding the slight deviations caused by the mutual action of the bow and arrow before explained —these in fact, as has been already shewn, to a great extent counteracting each other. Just as, for example, a hoop when in rapid motion may be slightly struck at the side, and a deviation from its path caused, which it nevertheless immediately recovers from, and continues in its original course.

The recoil of the bow, besides the motion in direction of aim, impresses a rotary motion upon the arrow about its centre of gravity. This tendency, however, to rotate about an axis through its centre of gravity is counteracted by the feathers. For, suppose the arrow to be shot off with a slight rotary motion about a vertical axis, in a short time its point will deviate to the left of the plane of projection, and the centre of gravity will be tne only point which continues in that plane. The feathers of the arrow will now be turned to the right of the same plane, and the velocity of the arrow will cause a considerable resistance of the air against them. This resistance will twist the arrow until the point comes to the right of the plane of projection, when it will begin to turn the arrow the contrary way. Thus, through the agency of the feathers, the deviation of the point from the plane of projection is confined within very narrow limits indeed.

A rotation about a horizontal axis would be prevented in the same way by this action of the feathers. Both these tendencies may be distinctly observed in the actual motion of the arrow.

If the foregoing reasoning be carefully considered, it will be at once seen how prejudicial to the flight of the arrow in the direction of aim any variation in the shape of that part of it in contact with the bow must necessarily be . for by this means a new force is introduced into the elements of its flight. Take for example the *chested* arrow, which is smallest at the point and largest at the feathers. Here there is, during its whole passage over the bow, a constant and increasing deviation to the left of the direction of aim, caused by the arrow's shape, independent of, and in addition to, a deviation in the like direction, caused by the retention of the nock upon the string. Thus this arrow has greater difficulty in recovering its first initial direction, the forces opposed to its doing so being so much increased. Accordingly, in practice, the *chested* arrow has always a tendency to fly to the left.

And so as regards the *bobtailed* arrow, which is largest at the point and smallest at the feathers, the converse of this is true. For here the tendency during its whole passage over the bow is to the right of the direction of aim, only restained by the retention of its nock on the string. But, as I have previously shown, the blow of the bow, during the last half of the arrow's passage, causing deviation also to the right, and in a degree, at least, to counteract the action of the string, there is a preponderance of deviation to the right for the arrow to overcome, in order to recover its initial direction: accordingly in practice, the *bobtailed* arrow has invariably a tendency to fly to the right.

Oddly enough, however, the bobtailed arrow has been looked upon as the easiest to shoot straight with, its shape having been considered partially to counteract the deviation to the left, believed to be caused by the action of the string. But as has been already shown, this left-hand deviation has no practical existence; the right-hand tendency, therefore, of the bobtail is an unmitigated evil. Excepting, indeed, to those who erroneously draw the arrow to the right of the eye; as, when this is the case, the arrow when pointing to the left of the mark, is apparently to the shooter directed straight towards it; thus the fault of the arrow operating in the opposite direction partially, counteracts the fault of the shooter. It is a bad system, however, to mend one fault by another. Better far to get rid of both.

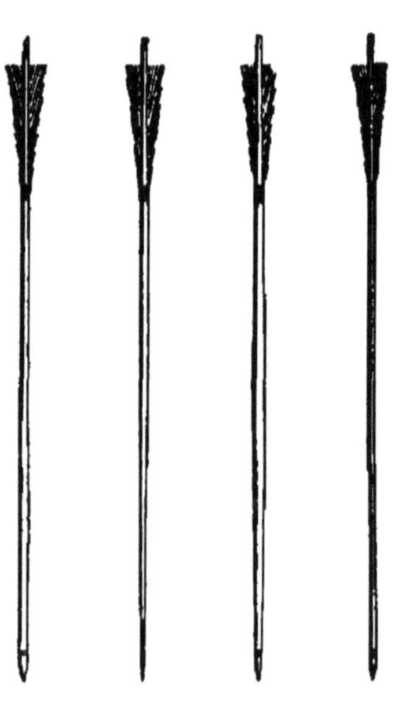

Bobtail. Chested. Barrell'd. Straight.

There is another arrow very much in use, called the *barrelled* arrow. This arrow is largest in the centre, and tapers thence to

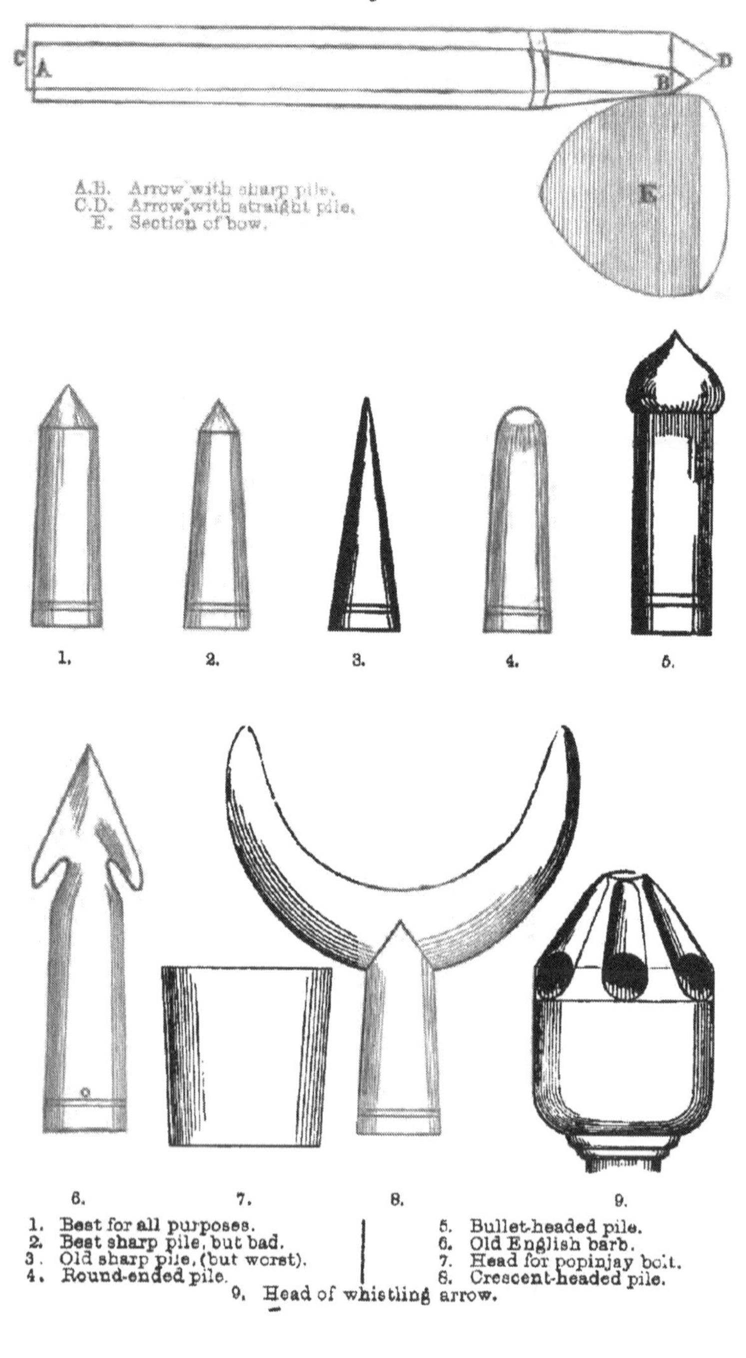

1. Best for all purposes.
2. Best sharp pile, but bad.
3. Old sharp pile, (but worst).
4. Round-ended pile.
5. Bullet-headed pile.
6. Old English barb.
7. Head for popinjay bolt.
8. Crescent-headed pile.
9. Head of whistling arrow.

both ends; it has a rapid flight, but does not follow the point well; and is additionally objectionable as a departure from the straight line. In short, it may be set down as an incontrovertible position in target shooting, that any shape of arrow that causes the centre of its thickness to vary in its relation to the edge of the bow, is radically bad. Therefore none other than the perfectly straight arrow is here recommended.

The *feathering* of the arrow is the most delicate part of the Fletcher's art, and requires great care and experience to effect it as it should be effected. Rather full-sized feathers are to be preferred, as giving a steadiness and solidity, as it were, to the flight; they should have a fair amount of *rib*, for if pared too fine their lasting qualities are diminished; and all three should be of the same wing, right or left. The turkey supplies most of the feathers now in use; those of the eagle and peacock, though most excellent, being too scarce to be generally attainable. The feathers of the "grey goose wing," so much spoken of in the legends of our forefathers, as guiding their unerring shafts to the heart of knight and yeoman, despite of "Milan steel" and "Jerkins buff," are now quite out of fashion; but as, of course, it would be absurd to suppose that they were not wiser in their generation than we are in ours, we must conclude that either turkeys did not exist in those days, or that geese have degenerated!

The *pile* or *point* is a very important part of the arrow. Of the different shapes in use, the blunt or square-shouldered pile is the only good one. In every respect, even for distant shooting, it is superior to all others; but the greatest advantage it possesses is, that if the arrow be overdrawn, so as to bring the pile on to the bow, it will not alter the direction of its flight, as is the case with all the sharp piles. (See plate 5.)

No. 1 (in the same plate) is the only one recommended for

target shooting. No. 5 has great penetrating power, for if it passes through the object struck, the whole "stele" will follow it, No. 6 is the old English barb. No. 8 is probably the shape of the pile used by the Emperor Commodus, who is said to have cut off the heads of ostriches at full speed! No. 9 is the whistling head, supposed to have been used to give alarm at night.

To prevent the pile coming off, either by damp or by a blow, it may be slightly indented on opposite sides by a gentle rap with a pointed instrument; a broken bradawl filed to a point is as good as anything.

The *nock* should be full and strong, and the notch as deep as will hold the string safely. To provide against the risk of splitting, I have found it a good plan to drill a hole through the solid part of the nock, as near the surface on which the string rests as may be, and to insert a piece of copper wire, which, when clenched or flattened at both ends, forms a safe rivet. A small Archimedean screw drill is the best for this purpose, but great care is required in using it, or it will cause the very evil it is intended to guard against.

As regards the *length* of the arrow, no arbitrary rule can be laid down: every archer must suit himself according to the length of his pull: hereafter I hope to lay down some principles which will guide him in this: for the present it will merely be observed, that no man's arrow, whatever his pull may be, should be less than twenty-six inches in length, as experiment has proved that a short arrow flies less steadily than a longer one. It is not absolutely necessary, though it is better, that he should pull the whole length of the arrow, provided his draw be *always to the same mark*.

The *weight* of an arrow must, to a certain extent, be regulated by its length, and the strength of the bow with which it is to be used; for if an arrow be a long one, it must have bulk sufficient to insure stiffness, and stiffness in proportion to the strength of the bow; 4*s*. 3*d*. for the lowest, and 5*s*. 6*d*. for the highest weight, are two extremes, within which every length of arrow and strength of bow may be properly fitted, so far as gentlemen are concerned. For ladies, 2*s*. 6*d*. and 3*s*. 6*d*. should perhaps be the limits. It must be borne in mind that a light arrow is a decided mistake for target shooting. Even flight arrows need not be less than 4*s*. in weight.

To preserve the feathers from damp, let a coat of oil paint be laid on between and for ¼th inch above and below them, and let this be afterwards varnished with a mixture of mastic and gold size, taking care that the *rib* of the feather be well covered, otherwise the desired purpose will not be attained. If the feathers be laid or ruffled by wet, they may be restored to their proper shape and firmness by being held for a short time before a fire, and kept turning, to prevent scorching.

Mr. Roberts mentions, and I have proved, a curious effect which is produced by feathering a light arrow at both ends, the wood being lightest at the pile-end; and the feather trimmed low at the nock and high at the pile-end; this, if shot against a wind, will return back again, like a Bomerang. If the same shaped arrow be feathered in the middle only, it will, in its flight, make a right angle, and no power of bow can send it any distance.

As the elevation should be regulated by the rise or fall of the left arm, and not by the weight of the arrow, the use of the same shafts at all distances is strongly recommended. Indeed, it is a great mistake to change any part of the tackle, bow or arrow, during the shooting, excepting in extraordinary cases; seldom, indeed, is the scoring bettered by such means.

Three arrows are usually shot at one time, and a fourth kept in reserve, in case of accidents. Now let it be remembered, that if the slightest variation either in shape or weight occurs amongst them, the line or the elevation is sure to be effected, to the serious detriment of accurate hitting; therefore too much care cannot be taken in their choice. Whatever kind or weight is used, let all the four be precisely similar in every respect.

Whether for store or daily use, the arrows should be kept in a quiver or case, made on such a plan that each shall have its separate cell, and so be insured from warping, or from having the feathers crushed. It is too much the custom to squeeze a quantity of arrows into a small quiver; let not the archer who prizes his tackle be guilty of this folly. They will wear out quite speedily enough, without the addition of ill-usage to hasten it. In drawing them from the ground, or the target, let the hand take hold as near the pile end as possible. Every archer should have an appropriate mark painted on each of his arrows, so that they may be easily distinguished from those of his neighbour.

It is a great point to have the arrow well stopped—that is, the wood should completely fill the pile, which otherwise, in striking against any hard substance, is apt to be driven down the stele, to the great detriment of the arrow, and often the destruction, by splitting, of the pile itself.

Chapter VI.

Of the String, the Bracer, and Shooting Glove.

THE STRING.

Of the bowstring very little need be said. The only good ones are of foreign make, and the very best are, I have understood, the produce of one particular maker, a Belgian, in whose family the secret of their manufacture is preserved with such jealousy as to cause a fear of its being lost, inasmuch as its present possessor is the last of his race.

A thick string is generally supposed to cast the steadiest, a thin one the sharpest; but, though preferring the latter myself, I have not been able to discover much practical difference between them; the strength of the bow must, however, somewhat regulate its substance. In any case the string should be round and even, with a tolerably thick eye at one end for the upper horn, and plenty of substance in the twist at the other to form the loop for the lower end of the bow. This loop is formed by giving the appropriate end of the string one turn round itself, and interlacing or twisting it three or four times afterwards; taking care to do this evenly and firmly, so as to prevent slipping, and waxing the end before doing so. The length of the string between the loop and the eye must of course be regulated by the

The Loop.

length of the bow; and ought to be such that, when the latter is strung, a space of at least *six* inches for a man's, and *five* inches for a lady's bow, should exist between the string and the centre of the bow.

The string for one inch above and five inches below the nocking point must be lapped with thread or thin twine, well waxed, of such a substance as nearly to fill the nock of the arrow, and this again, as far as is covered by the fingers when drawing, with a lapping of floss silk. The object of the latter is to render the loose smooth and even, and to supply the place of greese wholly or in part. Any substance that is of the right thickness, and at the same time smooth and even, may supply the place of the thread and floss, and will equally well attain the desired object. A piece of smooth vellum, or thin strip of buffalo hide, have been found to answer admirably, being pleasant to loose on and very durable. Whatever is used, the nocking point must be of just such thickness as will fill the nock of the arrow without splitting it; this is one of those minutiæ of archery essential to good shooting. If the string become frayed, it may be rubbed with beeswax, or thin glue; but if it be worn in any part, especially at the nocking point or the lower horn, let it be instantly rejected and replaced with a new one; for it is poor economy to risk the loss of perhaps a favourite bow, worth many pounds, for the sake of eighteenpence, the price of a new string.

A few spare strings should be kept in stock, free from damp; if in a tin case, so much the better.

THE BRACER.

The object of the bracer or armguard, it is almost needless to say, is to protect the left arm from the blow of the string, in the event of this striking upon it when loosed. By the expression "in the event of," it is especially meant to imply that no need exists for the string's striking the arm at all; and hereafter, and in its proper place, I hope to demonstrate that when this does occur, it is a fault to be amended, not a habit to be indulged in; and that if it *habitually* takes place, an insuperable barrier is thereby presented to certain and accurate shooting. Ascham indeed will have it that the bracer serves also a second purpose, viz., "that the string, gliding sharply and quickly off it, may make a sharper shoot,"—which would appear to be about as probable a result as would be the accelerating of a racer's speed, when in full career, by striking against a brick wall or any other obstruction. It is only, however, just to say, that he recommends the bow being so much strung up that the string shall avoid touching the arm at all; by which it may be concluded, that he merely meant to assert that the arrow's sharpness of flight was less injured by the string's striking against a hard smooth substance, such as that of which bracer's are usually composed, than against a soft yielding one, as the sleeve of the coat, for instance, would be—and not that there was any peculiar quality in the armguard of increasing, actually and positively, the cast of the bow. The bracer then is simply a protection to the arm in case of need—unless indeed, as regards the fairer portion of humanity, it may be said to be of service in confining and rendering harmless those pendant armlets of lace, crochet, and frippery that nowadays adorn their dresses, and the which, however serviceable in rendering more fatal the ethereal shafts of Cupid, are anything but conducive to the correct flight of those grosser and more material missiles that obtain in an archery field.

Too much care cannot be taken to see that, when fastened, no edge or corner protrudes that can by possibility obstruct the free passage of the string. I remember, upon one occasion some few years back and in my earlier days of practice, missing fifty-eight shots in succession, and only discovering at the end of that time that this mischance was entirely owing to one of the buckles having become loose, and so allowing the upper edge of the bracer at the time the arm was straightened to project some quarter of an inch beyond it; the string, in its passage back to the starting point, grazing against this projection, was for an instant arrested or else thrown out of the proper line; and thus the arrows either left the bow before the proper moment, and so fell short, or, receiving the same eccentric direction as the string, were cast about the field in every direction but the right one. Upon remedying this defect my shooting resumed its ordinary course.

In spite of good Ascham's "sharper shoot," a bracer made of moderately soft leather is preferable to a very hard one, as in this latter case the string on striking receives a greater rebound and vibration, which more or less injuriously affects the flight of the arrow.

The bracer must not be buckled too tightly on the arm, as, besides the discomfort and inconvenience this will occasion, it will serve to impede the free play of the muscles, and thus tend to destroy the accuracy of the shooting. It is a very good plan to have the upper edge of the bracer shaved thin, and then sewn on to the shooting coat (being still buckled as before), as this effectually prevents such an accident as that already related, and insures its fitting closely and tightly to the arm. The straps should be of such a length that the buckle shall be quite at the back of the arm, and not at the lower side, as in the latter case the sharp end will be in the way of the string—a very common occurrence by-the-bye. Instead of straps an elastic band with hook and eye has been lately introduced, and with every success.

Punch's advice to persons "about to marry" will doubtless be in the remembrance of many of my readers; it was comprised in the single word "Don't." I shall conclude the subject of the bracer with the same piece of advice addressed to those who are constantly in the habit of striking it, and about to do it again—"Don't."

THE SHOOTING GLOVE.

By this is understood the protection that every archer more or less requires to the drawing fingers of the right hand. It is the one accoutrement of the archer that requires perhaps more care and attention than any other; for, as it is certain that skilful and scientific shooting depends in great part upon an even, certain, and unvarying loose, so I have found such a loose is only to be attained by the help of the most perfectly fitting and accurarely adjusted shooting glove. As will be shown hereafter in treating of that part of archery (the loose), the great thing is to have the perfect command of the string, of the exact "how" and the "when" it shall be allowed to quit the fingers. This becomes almost an impossibility should the shooting glove be either too tight or too loose; in either case this necessary command is lost; in the one by the hold of the string (from the slipping of the glove) being insecure; in the other, by the fingers becoming cramped, and, so to speak, comparatively *hors de combat*. Again, too thick a glove prevents the proper "feel" of the string; too thin a one hurts the fingers, and causes them to flinch from the proper degree of sharpness required for the loose. And, once more, with too hard a glove the string cannot be with certainty retained till the proper instant of loosing; with too soft a one it is apt to get so imbedded as to require an unnatural jerk to be got rid of at all.

From all this it will be seen to be impossible to lay down any defined and specific rules for its size, shape, make, &c., &c., each individual requiring to be suited according to the peculiar nature of his own fingers, be they hard or tender, fleshy or otherwise; and it is therefore strongly recommended to every archer to be the manufacturer of his own shooting glove, as no other can fit him with the nicety and accuracy positively required. It may, however, be said generally that the thinner the leather composing it be, the better (provided always it be thick enough to protect the fingers from pain), as also that it be not so constructed as in the slightest degree to confine the hand or cramp the knuckles. A small piece of quill placed inside will also be found of material assistance in giving a clean and certain loose.

I must candidly confess, however, that the endeavours of ten years have hardly succeeded in producing finger stalls perfectly to my satisfaction. The following is one of several good plans for making them:—Let the finger-guards be made of a smooth, pliable piece of leather, perfectly independent of each other, and fastened behind with vulcanised India-rubber, and further kept in their places by rings of the like material passed over the middle joint of each finger; such ring having a thin tongue (also of India-rubber), about an inch or an inch and a half in length, fastened to the leather stall inside the hand. The hand is thus perfectly free and unrestrained, and the elasticity of the India-rubber prevents any tightness of the stalls

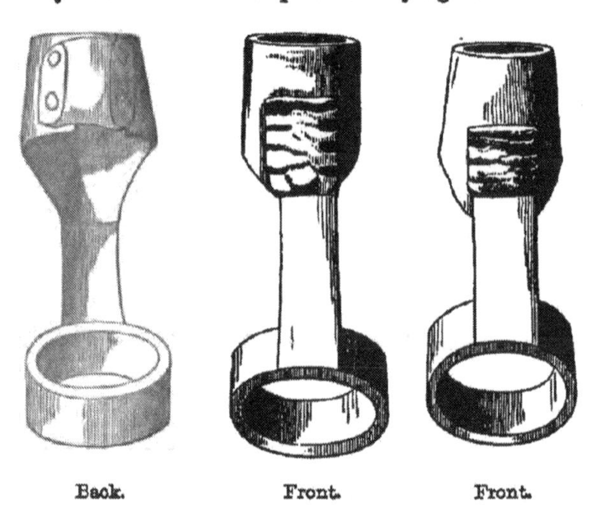

Back. Front. Front.

or confinement of the fingers; a guard or stop is placed upon each stall about half an inch from the top, by which the line of the fingers and position of the string is so regulated as to render the loose always uniform. (See accompanying sketch.) The merit of the vulcanised India-rubber is, as far as I know, due to Mr. Mason. Some have objected to the India-rubber, as stopping the circulation, and so numbing the fingers. This, I think, proceeds from having it too strong and tight. Latterly, instead of elastic rings, I have used silver ones, with good effect. If a little glue or resin be rubbed inside the stalls, it will keep them tight to the fingers, and the India-rubber back may be dispensed with.

Several excellent gloves have lately been brought out by Mr. Buchanan, of Piccadilly, very similar to those just described. For those who cannot bear the exposure of the tips of the fingers to the friction of the string in escaping, the glove here sketched may be highly recommended.

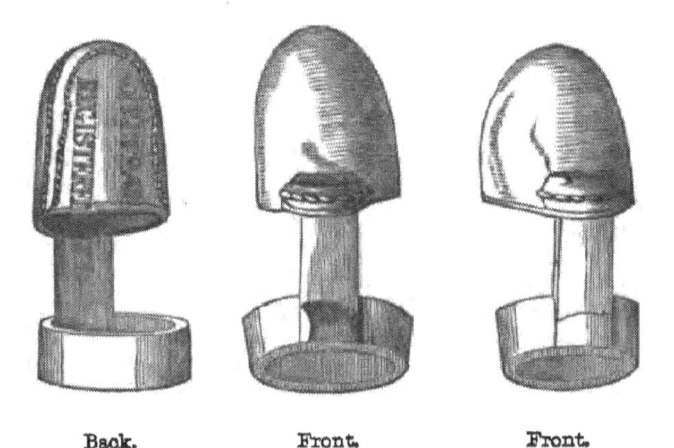

Back. Front. Front.

Mr. Buchanan's Glove.

There is a peculiar kind of finger-guard, known by the euphonious name of "tab," that requires some notice. It is simply a piece of flat leather lying inside the hand, and held in its place by the fingers being let into it at one end. It cannot, however, be compared with either of the gloves just described, either for evenness and certainty of loose, or for perfect command of the string. Still it must in fairness be stated that several excellent shots are in the habit of using it. This does not, however, alter my opinion as to its being decidedly an inferior method, as who shall say how much more these might have excelled had they adopted a different and more rational one?

Chapter VII.

Of the Grease Box, Tassell, Belt, etc.

THE GREASE BOX.

The grease box is an invention, as its name implies, for the purpose of holding grease, which may be lard, deer's fat, or any other "anti-sticking" mixture that the varying fancies of different individuals may delight in. There is nothing to be said against the use of such matters; but to those who, like myself, object to messing with grease, a lapping of floss silk, or any other smooth substance will be preferable, and will answer every required purpose: but this is one of those points that can be safely left to each individual Archer to decide, by his own experience, for himself, more or less assistance being required, according to varying strength and powers of the fingers. Thus much, however, should be said, that it is quite possible to have the string so slippery as to prevent that *perfect command of the time of loosing* which is a main ingredient in successful shooting. The grease box is generally made of wood, horn, or ivory.

THE TASSEL.

He must be a good Archer indeed who can do without this necessary appendage to his equipment. It is simply a tassel, made of green worsted, for the purpose of removing any dirt that may adhere to the arrow after it has been drawn from the ground. It need not be a yard in circumference, as, to judge from the stupendous size of their tassels, would seem to be the opinion of some Archers, but of as small a size as is compatible with its answering the required purpose.

THE BELT.

This is a strap with a small pouch attached, which fastens round the waist, and serves the purpose of holding the arrows, tassel, grease pot, spare string, and any other little paraphernalia of Archery, as also to assist in rendering one as hot and wretched as possible whilst shooting in warm weather. It is better to have a small deep pocket made in the right-hand side of the shooting coat; this will serve to hold the arrows, and the other matters can be hung on to a button of the coat, or, if it be liked better, kept in the ordinary pocket. These remarks, however, do not apply to the fair sex; the shooting belt is an evil they must perforce put up with.

Belt, &c.

THE SCORING APPARATUS.

The scoring apparatus is of various kinds and shapes, suitable to all tastes, and consists of a scoring card, a frame (generally of wood, silver, or ivory) to hold it, and a needle, enclosed, with the exception of the point, in a case of the like material, to prick the hits as they occur upon the card. It would be a waste of time and space to attempt a description of the different apparatus in use; suffice it to say that each Archer can suit his own taste, and, by having a plate struck for himself, can have the card arranged in that manner as will best accommodate the distances and number of arrows he is in the habit of shooting. For those who practise the national round, an excellent card and frame to suit, originated by Mr. Bramhall, of Norfolk, are in use. The former is so arranged as to embrace two rounds, and to keep a separate and distinct account of each distance and each individual arrow shot, and this all within a very small compass. The frame contains a pricker to mark with, and a pencil to do the additions afterwards. It can be strongly recommended to all those who take an interest in ascertaining the exact particulars of each day's shooting.

THE ASCHAM.

This term is applied to a small narrow cupboard, constructed for the purpose of holding the Archer's implements. It should be so arranged that the bows can stand upright, and each individual arrow in the same position, and sufficiently apart from its neighbour to prevent the feathers ruffling each other. The principal point to be mentioned, however, is not as respects the Ascham itself, but concerning the locality in which it should be kept. This should be in a room free from damp, and the temperature of which is as even as may be; if on the ground floor, to insure safety, the Ascham should be raised five or six inches from the ground. This is especially called for in country houses, as these are often built directly on the earth, and ill drained. The very best place for the Ascham is a room over the kitchen, as this gives that medium temperature, neither too hot nor too cold, which is especially suited to the preservation of bows, arrows and strings.

THE REGISTER.

This is simply a book, ruled and arranged in such a manner as to enable the Archer to keep an accurate account of his shooting. Those who have not been in the habit of having one can have no idea of the great interest with which it invests the most solitary practice, and how conducive it is to its steady and persevering continuance. It begets a great desire to improve, for no man likes to have evidence before his eyes of his pains and exertions being of no avail, and himself at a stand-still in any pursuit he takes an interest in; it ensures a due carefulness in the shooting of every arrow, since without it, the score will be bad, and therefore disagreeable to chronicle; it excites emulation, by enabling one man's average shooting to be compared with anothers, and restrains, by its sternly demonstrating figures, those flights of imagination occasionally indulged in by bad memories, as to feats performed and scores achieved. By noting too in this register the causes of failure at different times, a less chance will exist of their occuring again, as it keeps the same always in the mind's eye, and their necessary avoidance prominently before the attention. In short, the Archer will find the little trouble the keeping of it occasions him so abundantly repaid in a variety of ways, that having once commenced it, he will never afterwards be induced to abandon its use.

THE TARGETS.

The target is made of thrashed or unthrashed straw (rye is best), firmly bound together with tarred string, somewhat similar in its formation to a beehive, and is covered with stout canvass, upon which are painted five different coloured rings, white, black, blue, or inner white, red, and gold (commencing from the outside). It should be exactly four feet in diameter, neither more nor less, the breadth of all the rings being the same. This gives 9 3-5ths inches for the gold or centre, and 4 4-5ths inches for each of the rings to the right and left of it. The circles are valued as follows .—*Nine*, for the gold, *seven* for the red, *five* for the blue, *three* for the black, and *one* for the white. These figures, however, do not represent the correct value of the rings according to their respective areas; for reckoning the gold to score *nine* as above, strictly speaking the red should count but *three*, the inner white or blue *two*, the black *one and a quarter* (or five for every four hits), and the outer white *one*, (vide Waring's treatise, page 39). This incorrectness in the number scored for each ring, however, is altogether unimportant, for as one man's score is only good or bad as compared with another's, and all use the same target and mode of counting, each Archer gets the same proportionate benefit from the excess of counting, and so the comparative result is the same.

Formerly, if an arrow lodged on the border of two rings, the least valuable of the two was counted; but of late years the higher has been allowed—and this is right, as where a man hits two rings, he should have the option of choosing which he likes; of course, he will take the highest. I believe the Woodmen of Arden, however, still retain the old mode of reckoning the hits, allowing the Archer only to score the lowest of the two circles struck by the arrow.

The facing of the target should be covered with *nothing but paint*; there is too prevalent a custom amongst the target-makers of laying on, previously to the paint, a coating of whiting or some other villanous compound, in order to cheapen the process of colouring, and to smarten the appearance of the facing, and the consequence is, that after a day or two's use, and even without it, this original coating adheres to the arrow, and peals off in little flakes (of course carrying the paint with it), so that in a very short time, and long before either straw or canvass are one quarter destroyed, hardly a remnant of colour remains to distinguish the circles; and in addition to this, the Archer is bored with the necessity of removing this sticky compound from the end of his arrow, every time it is removed from the target. I know of nothing more annoying to the Archer than this. He has paid a high price for his targets, and perhaps for long carriage besides, and does not get a quarter of the wear out of the facings he has justly a right to expect. The proprietors of Archery warehouses, did they but consult their own interests, would soon put a stop to this obnoxious practice, the result of it being, that every Archer who has it in his power gets the facings of his targets made in his own neighbourhood under his own eye, instead of purchasing them. "Verbum sap." &c.

The present colours of the target are not well adapted for the most accurate shooting, being too bright and glaring, confusing the eye, and attracting it from the centre. Thus it is most difficult to aim at the gold, and not at the target generally. A black centre, with the rest of the target white, or *vice versâ*, would be much more conducive to central hitting. The rings might still be equally well marked.

THE STANDS.

These consist of three pieces of wood or iron, about six feet in length, fastened together by hinges at the top, and form a triangle, upon which the target is suspended. There is little or nothing to be said about them, excepting that, if made of iron, they should invariably be covered with a thick coat of leather or gutta percha; otherwise the arrows will be constantly broken against them, especially in windy weather; but even with this precaution, the wooden ones are the safest, and if these latter are faced from end to end with about two square inches of good solid stuffing of tow or shavings, inclosed in canvass, they will last a great many years, and never do injury to a single arrow.

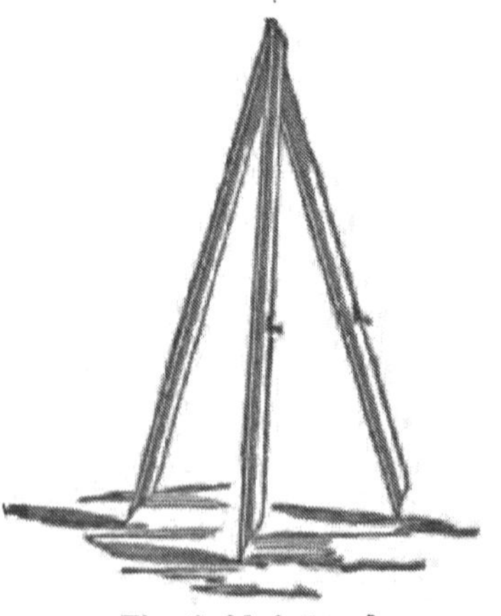

The Padded Stand.

A new kind of stand was introduced by the late Rev. J. Meyler, and is now, and has been for some years past, in use on the Toxopholite Grounds in Regent's Park. It is of iron, and constructed in such a manner that no part of the stand is visible to the shooter. Though considered perfection by its author, it is open to several objections; the chief of which are, that it is very expensive, can only be used as a fixture, and is occasionally the cause of a broken arrow; since, in spite of its careful construction, in dry weather the shaft will often rebound from the ground against it. I do not see, indeed, how the triangular wooden stand, well guarded, is to be surpassed.

The accompanying plate will give an idea of the Meyler stand.

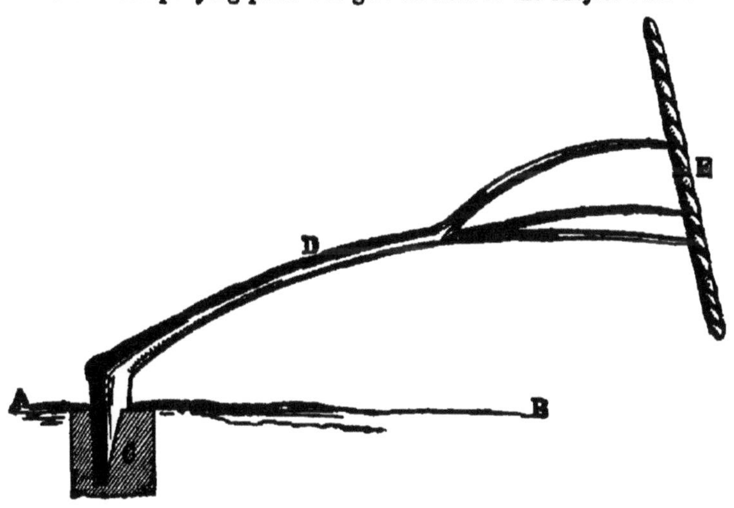

The Meyler Stand.
A.B Level of the Ground.—C. The Socket.—D The Stand.—E The Target.

THE QUIVER.

The quiver is commonly a case made of tin, to hold about a dozen arrows, sometimes having a small receptacle in the top to contain a spare string, a piece of wax, some twine and silk, and a file. These old-fashioned quivers are, however, very objectionable, as there is no provision made for keeping the arrows separate, so that too often they are squeezed in anyhow, and the feathers are crushed, and the arrows warped. The best sort of quiver (which now generally fits the travelling bow-case) is made of wood or tin, flat-sided, and fitted inside (at nine inches from the top and six inches from the bottom) with two shelves (one inch thick), bored with as many holes (half-inch in diameter) as there is room for arrows. These two shelves have the holes exactly perpendicular to each other, so that the arrows, passing through the two, are kept steady in their places, without danger of warping or crushing the feathers. If made of tin, the bottom should be lined with a piece of leather, gutta percha, or cork.

Chapter VIII.

Of Bracing and Nocking.

Different Modes of Bracing—Bend of the Bow—Ordinary Mode of Ascertaining its Correctness—Usual Direction as to Nocking—Its Objection—How to be Remedied—Position of Nocking Place—A Word of Warning.

In the previous chapters such plain directions, it is hoped, have been given concerning the various implements of Archery, as will enable each Archer to provide himself with the best of that kind his inclinations and means may lead him to adopt; and to avoid such as are in themselves radically bad, or likely to add to the difficulties he is sure to meet with before arriving at any great or satisfactory proficiency in the art. Having thus enabled him to form a choice as to his weapons, the next step is to endeavour to guide him in their use; and in the first place I shall notice a few minor matters, which, although of lesser importance in themselves, when compared with the more abstruse and difficult points connected with scientific Archery, yet must not on that account be altogether passed over in silence: and the first of these has regard to *bracing* the bow, which may be considered as the first preliminary operation in actual shooting. This is perhaps better known under the more modern appellation of *stringing*, and has reference to the act of bending the bow when unstrung, sufficiently to enable the shooter to slip the upper noose of the string into the nock. To effect this, three different modes have been practised. The first and most usual method is to set the lower horn of the bow on the ground (its back being towards the Archer) against the inside of the right foot, this being turned a little inward to prevent its slipping; then firmly grasping the handle with the right hand, and resting the lower part of the inside of the left hand upon the limb, just below the eye of the string, with a strong pull at the handle to bend the bow (the left hand and right foot forming the *points d'appui* of its two ends), the thumb and second joint of the forefinger of the left hand at the same time carrying the eye of the string into the nock. Novices, in first endeavouring to perform the operation of slipping the string into the nock almost invariably fail in doing so, but as invariably succeed in getting their fingers beween the bow and string; thus discovering that the string can do something more than discharge the arrow, namely, nearly cut their fingers off. To prevent this untoward result, I have here appended a sketch (from a photograph) of the proper position of the hand and fingers whilst stringing—expressly for their benefit.

Stringing the Bow.

The second mode is by identically the same action, excepting that the left hand takes the place of the right, and *vice versâ*. The third mode is performed by resting the lower horn of the bow upon the ground (the belly instead of the back being turned towards the Archer), and, whilst one hand presses the belly from the person, the inside of the other supports the upper end of the bow, and at the same time slips the string into the nock. Of this last mode of bracing, it may be briefly said that it is somewhat unusual, and seldom practised.

As regards the first two methods, opinions are divided; some, and I think the majority, advocating the grasp of the bow with the right hand, whilst the few maintain the left hand to be the best. It is, however, a matter so totally immaterial as hardly to be worth the slightest controversy; still, as Archers have made it a *vexata quæstio*, I may as well state the principal argument advanced by both sides in support of either proposition, and leave each to decide for himself afterwards as to which he likes the best to adopt. The advocates of the left-hand grasp, then, maintain that, as the bow when shooting is held with that hand, it should therefore be strung in like manner, as it saves the necessity of changing hands, and the action is more direct; whilst those who maintain the right-hand grasp, though they allow this, assert it is more than counterbalanced by the necessity the archer is under of turning his back upon the mark, or whatever or whoever else he is at the time fronting, if the bow be placed against the left foot instead of the right, which it must be if the grasp be with the left hand. But it may be said to every shooter, male or female,

Utrum horum mavis accipe.

To unbrace the bow the action is the same, with the exception that the string is slipped out of the nock, instead of into it. Either to brace or unbrace gracefully, and without effort, is an affair rather of knack, than of strength or force, and is therefore only to be learnt with a certain amount of practice.

The bow being braced, two things are to be carefully noted; firstly, that the bend be neither too high nor too low; and, secondly, that the string starts from both horns exactly in their centre, neither to the right hand nor to the left, but dividing the bow precisely in half from end to end; if this latter caution be not observed, the grain of the bow runs considerable risk of being unnaturally strained, and the bow itself of being pulled awry, and out of its proper shape, and sooner or later of breaking in consequence; it is even doubtful if the correct cast itself be not also more or less injuriously affected by any carelessness on this point. It is another of those minutiæ of Archery which is of more importance than might at first sight appear, and one that should always be attended to before the bow is allowed to discharge a single arrow. During a morning's shooting, too, attention should be occasionally directed to the string, to ascertain that the noose has not slipped a little awry, which it will sometimes unavoidably do. Concerning the first point, it has been already stated, when speaking of the string, that, as regards a man's bow, the distance from the inside of the handle to the string should not be less than six inches. The advantages of a lower or smaller bend than this are that the bow casts quicker and further (owing to the greater length the arrow is acted upon by the string), and that the wood is less strained, and in less consequent danger of breaking; but to be put against this are the facts that the cast is less steady, and the probability of striking the bracer before the extreme point of the string's recoil (already asserted to be fatal to accurate shooting) much greater. Individually, I prefer the high bend, as giving much greater steadiness, tending more to secure the correct flight of the arrow, and making the drawing of the bow easier (the distance to be pulled being less), and have never found the loss of cast or the danger of breaking sufficiently great to induce an alteration of that opinion. I should therefore recommend the bow's being strung up, *at the least*, six inches.

It has long been the custom, in order to ascertain the amount of bend of the bow, to place the fist perpendicularly upon the interior of the handle (at the centre of the bow), at the same time raising up the thumb as high as it will reach: should the string then just touch the extremity of the thumb, the bracing is probably there or thereabouts correct: if higher or lower than the thumb's extremity, it is probably too great or too little, as the case may be. This is not, however, an infallible test, as the size and length of the hands of different individuals vary materially; but each Archer can once for all ascertain how near his own hand, placed in the above way, marks the distance he prefers, and, bearing this always in mind, brace his bow thereby equally as well as if his hand marked it exactly.

I shall now proceed to the point of *nocking*, though, strictly speaking, the next in order should be that of *standing;* but, as this latter is so intimately connected with, as to be almost inseparable from, the subjects of *position, method of drawing, aiming, &c.*, which are decidedly after-matters, it will be included in these, instead of being separately treated of. Nocking is the most simple operation of Archery; the usual directions given for performing it are as follows:—"Holding the bow by the handle with the left hand, and turning it diagonally with the string upwards, with the right hand draw an arrow from the pouch, and grasping it about the middle, *pass the point under the string and over the bow;* then placing the thumb of the left hand over it, with the thumb and first finger of the right hand fix the arrow firmly on the string, the cock feather being uppermost." There is one objection, however, to that part of them which directs the shooter to "pass the arrow *under* the string"—an objection, curiously enough, entirely overlooked by all the authors upon Archery—and it is this, that by doing so, and owing to the somewhat intricate passage the arrow is made to traverse, the bow is very apt to become pitted by the point of the arrow, and in most Archers' hands who nock in this way speedily assumes the appearance of having had an attack of some mild species of measles or small-pox, to the great injury of the bow, both as regards beauty and safety, especially when made of yew; this most valuable wood of all being of a soft and tender character. It is true it may be argued that it is the Archer's own fault for not using proper care and attention whilst performing the operation; but during the excitement of matches, or in rapid shooting, this is no easy matter; and thus it will be generally found that the bows of Archers who nock in this way are more or less indented in the manner mentioned. Are the *important* points of Archery, too, not sufficiently numerous and difficult to bear constantly in mind without adding another to the list, unnecessary and altogether useless? If it had even the recommendation of being more easily, or more quickly performed, it would be something in its favour; but neither of these arguments can be advanced by its advocates, neither does it possess one single advantage to counterbalance its serious objection.

I cannot imagine a plan of *nocking* more simple and easy than the following:—The bow being held by the handle with the left hand, let the arrow be placed with the right *(over* the string, not *under)* upon that part of the bow upon which it is to lie; the thumb of the left hand, being then gently placed over it, will serve to hold it perfectly under command, and the fore-finger and thumb of the right hand can then take hold of the nock end of the arrow, and manipulate it with the most perfect ease in any manner that may be required. Five minutes' practice will be sufficient to render this mode of nocking familiar and easy to any Archer.

The nocking place should be exactly upon that part of the string which is opposite the spot of the bow over which the arrow passes—that is to say, the arrow when nocked must be precisely perpendicular to the bow. If either above or below this point, the arrow will not have a good flight; and should it happen to be above a trifle so either way, the safety of the bow is also compromised, and its cast injured. Care must be taken that the nocking part of the string *exactly* fills the nock of the arrow—it must be neither too tight nor too loose; if the first, it may, and probably will, split the nock; if the second, the shaft is apt to slip when in the act of drawing, and the correct elevation and its proper flight be lost thereby. The degree of tightness should be such, that the arrow, if nocked and allowed to hang, should just be retained by the string—that is to say, sufficient to support the weight of the arrow.

I must add a word of warning to the young Archer against that objectionable but too common plan of attempting to alter the range of the arrow by changing the nocking-point, making it higher or lower as they wish to increase or diminish it. For the reasons above given, a worse system cannot be adopted.

Chapter IX.

Of Position.

The Standing—Requirements of a Good Position—What to observe and what to Avoid—How to Grasp the Bow—Waring's Method—The Correct One—Whether the Bow should be held Perpendicularly or Otherwise.

Ascham has made *standing* the first of his well-known five points of archery; but, as the term appears a most insufficient one for including all that has to be said respecting the attitude and general bearing of the Archer whilst in the act of shooting, I have preferred the expression "position" as being more applicable and comprehensive. Under this head will be included, not only what may be considered as more particularly appertaining to it, namely, the footing (or standing according to Roger) and attitudes of the Archer (irrespective of what may more properly belong to the point of *drawing)*, but also the manner in which the hand should grasp the bow, as well as the exact position of the bow itself.

And, first, as to the footing or standing, and attitudes of the Archer. Concerning these, it may safely be asserted that as many varieties exist as there are Archers to give them existence. At any rate, certain it is that hardly any two shoot precisely in the same form, and very few without some individual mannerism, *suus cuique modus.* Such being the case, it would be venturing too far to assert that but *one* position is good, or any particular *one* the best (indeed, it is doubtful if any Archer could be instanced as having attained *perfection* in this respect); but, nevertheless, some general rules are necessary to be borne in mind, and can with confidence be laid down, in order to control such mannerisms, and restrain them within harmless limits. For numberless examples might be given, where the unfortunate body and limbs are twisted and contorted to such a degree, and made to go through such wonderful acrobatic evolutions, as not only to violate all the requirements of grace and elegance, but also most effectually to prevent the possibility of even moderate hitting. Such faults would appear to have been common in Ascham's time, as well as in our own, for he gives us many instances of them. None, however, will be quoted upon the present occasion; but it will be rather endeavoured to lay down such plain directions as may prevent the assumption of attitudes inimical to good shooting, the reader being left to his own common sense to avoid such as do violence to gracefulness, and are repulsive to the looker-on.

An Archer's general position, to be a good one, must be possessed of three qualities—namely, firmness, elasticity, and grace: firmness to resist the force, pressure, and recoil of the bow; for if there be any wavering or unsteadiness, the shot will probably prove a failure;—elasticity, to give free play to the muscles, and the needful command over them, which will not be the case should the position be too stiff;—and grace, to render the shooter and his performance an agreeable object to the eye of the spectator. It so far fortunately happens that the third requirement, namely that of grace, is almost the necessary consequent of the possession of the other two; for the best position for practical results is almost sure to be the most graceful one. At any rate, experience proves that an awkward and ungainly style of shooting is seldom or never successful. Bearing in mind, then, the above three requisites, I shall endeavour to discuss what is and what is not the best position for combining them.

The first point that calls for remark is the footing or standing, and to this part of "position" there is little or nothing to be added to what has already been recommended in other treatises on Archery. The heels should be about six or eight inches apart, not further; for it is neither necessary nor elegant for the shooter to straddle his legs abroad, and look as if he were preparing to withstand the blow of a battering-ram, whatever his feelings upon the subject may be. The feet must be flat and firm on the ground, both equally inclining outwards from the heels, so that the toes be some six or seven inches wider apart than they; the position of the feet as regards the target being such, that a straight line drawn from it would intersect both heels—that is to say, the standing must be at right angles with the mark.—(Vide Frontispiece.)—The knees must be perfectly straight, not bent in the slightest degree. Some Archers violate this rule; but could they once see themselves, or understand the ludicrous-looking object they present to the spectator by doing so, they would hardly be tempted to continue the practice. The weight of the body should be thrown equally on both legs; for, as Mr. Roberts very justly observes, a partial bearing on one leg more than on the other, tends to render the shooter unsteady, and enervates his whole action. In short, the footing must be firm, yet at the same time easy and springy, and the more natural it is the more likely it is to possess these qualities.

If the foregoing rules respecting the footing be accurately observed, it will be found that the side only of the Archer's person is turned towards the target; and this is what has been invariably recommended by every author upon Archery, and is indeed the proper attitude. The left shoulder must not, however, be additionally forced forward, set in a vice as it were, but allowed to maintain its natural position—otherwise the required element of elasticity will be lost. The body should be upright, but not stiff; the whole person well balanced; and the face turned round, so as to be nearly fronting the target, with the expression calm, yet determined and confident—for nothing is more unsightly than to see the "human face divine" distorted by frowning, winking, sticking out the tongue, and the like—the whole attitude, in short, should be generally suggestive of power, command over the muscles, and the *will* to use them so as to produce the desired result.

During the brief period of time between the assumption of the footing and the loosing of the arrow, some slight alteration of the body's attitude first assumed will of necessity take place. During the act of drawing and aiming, the right shoulder will naturally come a little forward, and the left shoulder retire a little backwards. Indeed, were it not so, the shooter would be the very personification of awkwardness. The slightest possible inclination forward should also be given to the head and chest. The object of this is to bring the muscles of the chest into play to assist those of the arms, and is what good Bishop Latimer called "laying the body in the bow."

Not stooping, nor yet standing straight upright.

As Nicholl's "London Artillery" hath it.

A great many Archers bend the body very considerably from the waist; but this is most highly objectionable on every account. There is nothing to be said for it, and everything against it. Indeed, the shooter who adopts this position requires so much of his wits and muscles to keep himself from tumbling on his nose, as to have but little of either left to enable him to hit the mark. Not that he is to run into the opposite extreme, and look as if he had a ramrod down his backbone, or was without vertebræ at all; but the same rules apply to this point as to every other connected with the *modus operandi* of shooting, namely, that any strained or unnatural attitudes are not only ungainly and awkward, but also highly prejudicial to the success of the shooter. A warning must likewise be given against bending the head too much forwards. This, however, brings with it, fortunately, its own speedy punishment, for when it takes place, the string, in recoiling, will every now and then give the unfortunate Archer such a merciless rap upon the nose as effectually to cure him of the fault,—for the time being at all events; for Archers who have once experienced the penalty of this mistake, will not be at all inclined to undergo a repetition of it, if it can by any possibility be avoided.

I shall now proceed to the second part of my subject—which is indeed a most important one,—namely, the manner in which the hand should grasp the bow whilst in the act of shooting, and the exact position of the bow itself; that is, whether this should be perpendicular, or more or less oblique.

As regards the first matter, namely, the manner in which the hand should grasp the bow, it may be once more asserted that the most natural and easy position is also the best; in fact, this remark is applicable to almost every point connected with Archery, and cannot be too much and too often insisted on. Should the wrist and hand, then, be any way unnaturally employed, bad results immediately ensue. For example, should the grasp be such as to throw the fulcrum much below the centre of the bow, the lower limb runs great risk of being pulled awry and out of shape, which sooner or later will cause it to chrysal and break. And, again, if, as Waring and others inculcate, and too many follow, the wrist be "turned in as much as possible," the left arm must, perforce, be held in such a manner, and in so straightened a position, that not only will it present a constantly recurring obstacle and diverting influence

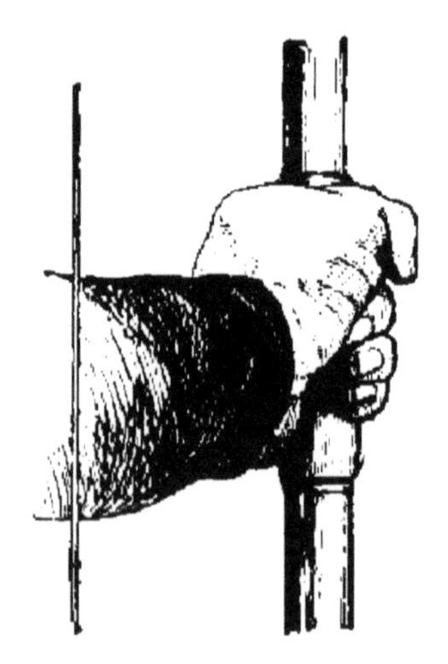

Waring's Method.—Wrong

to the free passage of the string, but also be the cause of an increased strain and additional effort to the shooter himself, besides taking all spring and elasticity out of him. If the reverse of this method be adopted, and the wrist be turned intentionally outwards, as some do (by the by, this is rather a peculiarity of the fairer sex), the whole force of the bow is then thrown entirely upon it, and it becomes unequal to the task of sustaining its pressure and recoil: thus, as in every other instance, extremes are bad, and to be avoided.

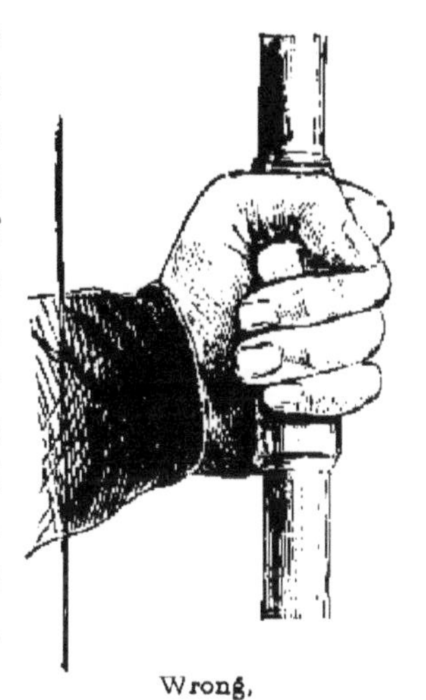

Wrong.

When the arrow is nocked and the footing taken, let the bow lie easily and lightly in the left hand, the wrist being turned neither inwards nor outwards, but allowed to remain in that position that nature intended for it; as the drawing of the bow commences, the grasp will intuitively tighten, and by the time the arrow is drawn to the head, the position of hand and wrist will be such as to be easiest for the shooter, and best for the success of his shot.

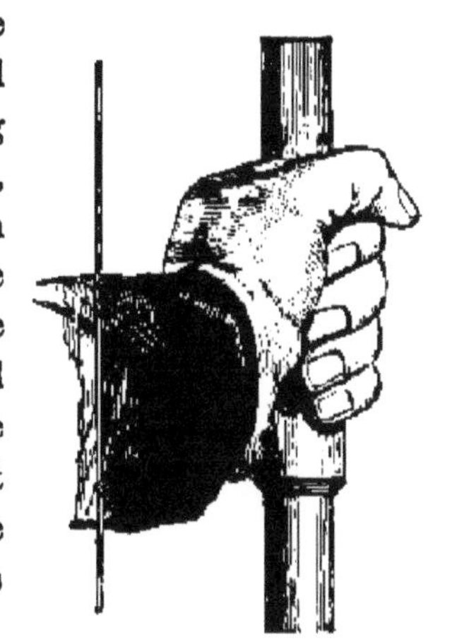

Some Archers have a habit of letting the thumb of the left hand lie extended along the belly of the bow, whilst others extend the forefinger, apparently to keep the arrow in its place. Both these habits are bad, as tending to weaken and unsteady the grasp, and as causing the jar of the bow to be more sensibly felt.

Regarding the position of the bow, whether it should be held, when drawn up for the aim, perpendicularly or more or less obliquely, opinions are pretty equally divided—the preponderance being, perhaps, rather to the side of the latter. I think, however, that sufficient reasons can be adduced as to leave no doubt that the oblique is the better method. For, firstly, the bow comes a little to that position naturally, the wrist requiring a slight twist to hold it quite perpendicularly; secondly, in a side wind blowing towards the face of the Archer, the arrow is more easily retained on the bow; and, thirdly, it gives the elbow of the left arm a slight inclination outwards, which is so far advantageous as assisting to keep that arm out of the way of the string. I know of no advantage possessed by the perpendicular holding of the bow to counterbalance the above advantages appertaining to the oblique. It is therefore recommended that the bow be held somewhat in the latter direction.

Chapter X.

Of Drawing.

Drawing an Essential Feature—Example of Bad Methods—Modes Adopted by Different Archers—The Best System—Inability of Devices to give Certainty of Draw—Position of Left Arm a most Important Feature—Mr. Waring's Arm-Striking Theory Denounced—The Necessity of the Unobstructed Passage of the String Demonstrated—Proper Position for the Left Arm—the Length of the Draw.

Whether Ascham's assertion that "drawing is the better part of shooting" be strictly correct or not, one thing is certain, that at any rate it forms one of its most important features; and upon the manner in which it is accomplished very much depends, not only the ease and grace of the entire performance, but the accuracy and certainty of the hitting. Now, though it is not asserted that but one method of drawing exists, whereby a man may attain to great scoring, it is nevertheless maintained that there are many modes in common practice at the present day, by the use of which such scoring is effectually prevented. A small volume might be written in describing the different *bad* methods of action adopted by various Archers to accomplish this part of shooting. One I have seen took first a deliberate aim at his own toe, then an equally careful one at the sky above his head, and finally at his mark; it is perhaps needless to add that he seldom or never hit it. Another was wont to go through the most extraordinary gyrations with both arms, moving them about somewhat like the sails of a windmill during the whole process of drawing until the very moment of the arrow's departure—where to, until it dropped, neither he himself nor any of the lookers-on had the remotest chance of divining. Several make a sort of see-saw of the bow and arrow, drawing the latter backwards and forwards for the last few inches, till the ill-treated weapons are at last allowed to separate. Such tricks as these, and many others like them too numerous for description, are methods of drawing that *do* prevent good shooting, and none are or can be correct that in so glaring a manner do violence to gracefulness, or the first principles of common sense. But, putting aside such eccentric performances as these, there still remain several different methods of drawing, that may fairly admit of discussion as to their respective merits, and these I shall proceed to notice.

Some Archers, and good ones too, extend the bow-arm fully and take their aim before they commence drawing at all. I cannot, however, think that this method is to be commended, as it has an awkward appearance from the necessity that exists of stretching the right arm so far across the body in order to reach the string, and materially increases the exertion necessary to pull the bow. The same objections apply, though in a less degree, to drawing the bow a few inches only, and then extending the arm and taking the aim. A third method to be noticed, the very opposite of that described, is, when the arm is extended, and the arrow drawn home before the aim is attempted to be taken at all. This, at the first view, has apparently a great point in its favour, namely, that it insures the arrow's always being drawn to the same point; but is objectionable, nevertheless, as being most trying both to arms and bow, as being generally ineffective, not particularly graceful, and causing the proper loose to be constantly missed, from the great overstrain that is laid upon the drawing fingers of the right hand. Another method is, to make the pulling of the bow and the extension of the left arm a simultaneous movement, and to such an extent, that the arrow shall be *at the least* three-fourths drawn at the time it is brought upon the aim—the right arm being at this time so much raised, that its elbow shall be on the same level as the drawing fingers. This is the system adopted by the generality of good Archers, and is decidedly the best, as being the most graceful in action, and by far the easiest as regards the pulling of the bow. There is some difference of opinion amongst those who adopt this plan, as to whether the arms and bow should be brought to the point of aim from beneath that point, or brought round and above, and then lowered to it (in either case, whether upwards or downwards, in a perpendicular line), or whether this should be done by a horizontal motion. The first method appears to me to be the simplest and most direct, since the drawing most naturally commences from beneath the point of aim, and it seems rather going out of the way to make an upward circular motion in order to get above it, for the sole purpose of again descending to it. As regards the horizontal movement, it is objectionable, as having a tendency to carry the arm across the target, and so out of the true line.

At this point (the arrow being at the least three-fourths drawn and the aim found), a further matter for discussion amongst Archers is, whether the continuation of the pull to its finish should be immediate and without pause, or otherwise; that is to say, whether the entire drawing should be one continuous act, from the first moment of pulling and raising the bow to the loosing; or whether the arrow should be held quiescent for a short time after the aim is found, so as to steady and correct it, and then be drawn to the loose. Ascham maintains that the first is the only correct method, and calls the second a shift; but a very great deal of experiment has proved that very little, if any, advantage is possessed by either system over the other, and that the Archer may adopt whichever he pleases without detriment to his shooting, provided only that the pause, if he make it, be a very slight one. My own predilection, from habit perhaps, is rather in favour of the continuous draw, and it is certainly somewhat less laborious, as if once a stop takes place, a renewed effort is required to complete the pull; but, upon the whole, the difference between the two methods is so trifling, that it may safely be left to the option of each shooter, as before stated, to choose for himself.

I shall therefore venture to recommend, as being, all things considered, the best system of drawing, that the pulling of the bow and the extension of the left arm be a simultaneous movement; that this be to the extent of drawing the arrow at the least three-fourths of its length before the aim be taken (if to such a distance that the wrist of the right hand come to about the level of the chin, so much the better); that the aim be found by a direct movement on to it from the starting-place of the draw; that the right elbow be well raised; and that the arrow be then pulled home, either with or without a pause, preference being rather given to the latter.

One of the main features of good drawing is, that the distance pulled be precisely the same every time, that is to say, the arrow always be drawn to identically the same spot. Unless this be accomplished, the elevation must be more or less uncertain, since the power taken out of the bow will, of course, be greater or less according to the extent it is pulled. A great many devices have been tried and practised to make exact similarity in the distance drawn a matter of certainty, by notching the extremity of the arrow for instance, so that the left hand may feel when it has reached a certain point, and by other contrivances of the like nature, and producing the same effect. But such devices never have a beneficial result; for when the eye and mind are fixed on the aim, concentrated upon it as it were (as they should be), if anything occurs to distract either, the shooting is sure to become uncertain and unequal. Some Archers endeavour to obtain a certain guide to the length of draw by means of the right hand, making this be felt in some particular part of the cheek. One who shot at one of the earlier Grand National Meetings actually held on by his own nose, the thumb being the instrument of fixture; another will put his tongue in his cheek, and hold on by that. The same objections apply to these as to the first-mentioned device; and they are additionally objectionable as being extremely unsightly and ungraceful, besides preventing the elasticity of finger required for a good loose. There appear, indeed, to be no artificial means by which similarity of draw can be beneficially obtained. Nothing but constant and unremitting practice will serve the Archer here.

The pile of the arrow should not be drawn on to the bow—at least it is better that it be not—as, unless it is exactly the same shape as the arrow itself, it will throw the latter out of the line. (See Plate 5.) Moreover, more or less danger will exist of the arrow's being pulled and set inside the bow, when such a smash will probably take place, as will be anything but soothing to the nerves of the shooter, or safe to his bow or eyes. This rule can be the more safely pressed on his attention, as there is no object gained by violating it. It is, therefore, recommended that the arrow be pulled just to that point where the commencement of the pile touches the bow and no further. Thus the arrow should be longer, by the length of the pile, than the Archer's actual draw.

All Archers, good, bad, and indifferent, are peculiarly subject (more or less) to one failing, namely, that of completing the draw, after the aim is taken, in a somewhat different line to that occupied by the arrow; instead of making it, as they should do, an exact continuation of that line; dropping the right hand, or letting it incline to the right, or both—the effect being to cast the arrow out of the direction it had indicated, and by means of which the aim had been calculated. Here, again, nothing but the most constant and untiring practice will serve the Archer; but his attention is most particularly directed to this most common failing, as it is one of which he will very often be entirely unconscious, though the cause of his continually missing his mark. The very best of Archers need to bear constantly in mind the necessary avoidance of this fault, for however skilful he may be, however experienced and practised a shot, he may be quite sure that it is one into which he will be constantly in danger of falling.

Now let it be remembered that the right hand must always be drawn to the same spot for all kinds of target shooting, be the distance what it may, and the arrow be pulled the same length. Some Archers have a very bad habit of varying the length of their draw at different distances, whilst others endeavour to accomplish the desired elevation by raising or depressing the right hand. This is all decidedly wrong. It is the left arm, and the left arm alone, that should do this part of the work, this being elevated or depressed according to circumstances, the right hand being maintained invariably in the same position at the moment of the arrow's departure. This is an incontrovertible rule in Archery to obtain a true elevation, and one that admits of no variation, however many Archers of the present day may be disposed to dispute its correctness.

A further and most important point, and one that applies to avery Archer, let his method of drawing be what it may, now calls for particular attention, namely, the *amount of extension to which the left arm should be subject* at the final completion of the draw, that is at the moment of the loose. Concerning this, it is without hesitation affirmed, that if the left arm be stretched out quite straight—"held as straight as possible," as is generally taught by all writers upon Archery, and more especially as already mentioned by Mr. Waring—accurate shooting at once becomes unattainable, owing to the difficulty, amounting almost to an impossibility, of keeping the string, when on the recoil, from habitually striking the bracer; or, in other words, that the constant striking of the string upon the bracer is inimical to certain hitting. This, at the first view, will strike many Archers as a most startling proposition, especially if their young ideas have been taught how to shoot from Mr. Waring's "Treatise on Archery, or the Art of Shooting;" as he not only recommends the direct reverse, namely, that the left arm *shall* be held quite straight, but clenches the matter by inculcating, in addition, that "it (the left arm) be so turned in, that the string strikes it when loosed." Now, attention is the more particularly directed to this, as it is believed, most anti-hitting instruction of Mr. Waring, as, from the large circulation his Work has had, and the reputation, somehow or other, enjoyed by himself as one of the good shots of his day (though, on reference to his scores, I find him to have been barely, if at all, more than third-rate, even at the low standard of shooting then prevailing), many Archers have carefully guided their practice by the rules he has laid down. Had he directed the shooter to stand on his head whilst drawing, or to shut his eyes whilst aiming, it had hardly been more injurious doctrine; indeed, much less so, as the self-evident absurdity of either would have prevented its ever being attempted. As it is, however, his arm-striking theory has kept many a promising Archer a slave in the dark regions of Muffdom, who otherwise might speedily have emerged from its precincts, and shown forth to the admiring gaze of the Archery world a full-blown Robin Hood.

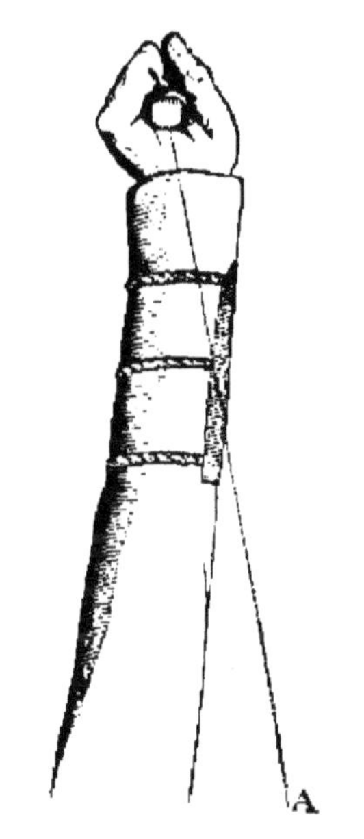

A Perpendicular Shot.
A.B.—The String's Course.

Before leaving Mr. Waring's Treatise, it should be observed, that he appears to have arrived at the conclusion that the left arm should be held in such a position that the string may always strike it, not because he actually thought the arm-striking beneficial as against the reverse, but as a guide to insure the left wrist being "turned in as much as possible;" he erroneously imagining that the bow could not be held with firmness in any other way. Certainly one system secured the other. I will now, however, endeavour to demonstrate the correctness of the proposition I have laid down, namely, that the constant striking of the string upon the bracer is fatal to good shooting.

Now, let it be borne in mind, that the flight and direction of the arrow are entirely caused and governed by the string's action upon it, the power of the latter being, of course, obtained by the recoil action of the bow, and that this government of the string over the arrow lasts to the extent of its force, in comparison with the other forces put into play by the act of shooting, during the entire passage of the former (after being loosed) from the extreme point of the draw to the like point of the recoil—at least, it should do so, as the arrow does not properly part company with the string before this passage is completed. This being the case, it follows that if, during the string's progress, any obstruction occurs to alter its natural and original direction, or to give it any vibratory or irregular motion, an immediate prejudicial effect of the like nature must be communicated to the arrow's flight; or the obstruction, if it be sufficiently direct to arrest or stay the string in its course even for an instant, must cause the arrow to leave the string before the latter has reached the extreme point of recoil, and thus, the proper fling of the bow not being communicated to it, the arrow must drop short.

Now, if the string strikes the bracer *previous* to its extreme point of recoil, it, of necessity, becomes subject to one or other of the two evil influences mentioned; for, as the bracer follows the line of the left arm, and the line of the left arm is altogether different from that traversed by the string from point to point, it becomes obvious that an alteration and irregularity of the string's line must take place over such portion of its passage as may exist between the spot where it strikes the bracer and the extreme point of its recoil, in addition to any vibratory or improper motion communicated to it by the blow itself. Thus the arrow, if it do not fall a victim to the second mentioned evil, namely, that of leaving the string too soon, must perforce become subject in a greater or less degree to the irregularities and misdirection mentioned as communicated to the string, and its accuracy of flight be entirely prevented thereby. It may, even if the amount of obstruction be insufficient to free the arrow, though enough to deaden the string's progress for the remainder of its passage, be subject to all the evil influences described. If the bracer be a hard one, the arrow will most probably lose little of its proper impetus (hence Ascham's "sharper shoot"), but simply be cast irregularly and out of its proper direction. If a soft one, it will probably be thrown short, though likely enough in its correct line. In short, it is abundantly clear, and the fact must commend itself to the reflection of every Archer, that, unless the string has a clear passage from end to end, the arrow can neither get the proper impetus from the bow, nor avoid receiving an eccentric momentum.

It is possible, however, that the string may strike the bracer, but do so only at the extreme point of its recoil, and not previous to it; or, in other words, there is just *one* spot where the string may strike the arm, without its becoming subject to any misdirecting or arresting influence. Now, could the Archer attain to such perfection of relative position in every respect, such precise similarity of drawing, &c., &c., as to ensure that the string, each time it was loosed should touch this spot, and this one only, he might indulge his penchant for arm-striking (if he have it) without, in all probability, any injury to the flight of his arrow or the accuracy of his shooting, though neither would be improved thereby, nor any conceivable advantage be gained by his doing so. Still he might, in such a case, revel in the indulgence of his crotchet, and be happy. Such required perfection of position, &c., &c., however is practically unattainable; and this being so, it results, that if the left arm be so straightened that the string *habitually* strike upon it when loosed; or if any other peculiarity of position or method of drawing bring about the same effect, some mischance, from the causes and of the nature already described, is constantly happening to the arrow, and marring the success of the shot; and this mischance will be of more or less importance, according as the distance between the part of the bracer struck, and the extreme point of the string's recoil be greater or less. *Crede experto.*

The Archer is, nevertheless, not to run into the opposite extreme, and deliberately bend his left arm, as in this case he will certainly, to a great extent, lose his power of resisting the force and pressure of the bow; but if the left arm be held out naturally and easily (the elbow being turned a little outwards and upwards), without a conscious effort either to straighten or to bend it, it will have just that position which will most easily enable him to withstand the force of the bow with firmness, and, at the same time, one that will allow of a free and unobstructed space for the passage of the string; especially if the left hand grasp the bow in the manner already advised in the chapter on "position."

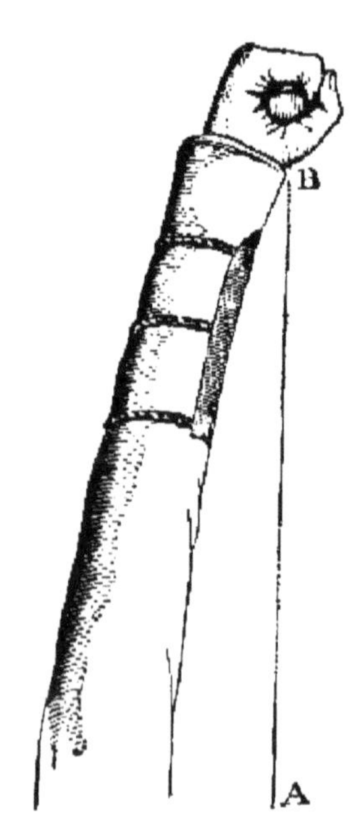

A Perpendicular Shot.
A.B.—The Course of the String.

The *length* of each Archer's draw must, of course, be regulated by the length of his arm, and should be such that, when the arrow is drawn to the loosing point, its nock should be in a perpendicular line with the right eye, the level of the arrow being a shade lower than that of the chin. (Vide Frontispiece.) A direction of pull higher than this is not recommended, as in this case, at the longer distances (at 100 yards, for instance), unless the bow shot with be a very strong one, or the arrows very light (which they should not be for target shooting), the proper sight of the mark is lost by the excessive but required elevation of the bow-hand and arm. For the reasons hereafter to be given, when treating of the "aim," the pull *to* the ear is decidedly rejected. The Archer need be under no apprehension that, in the way recommended, he will be unable to pull a sufficient length of arrow; if a man of six feet, he will with ease be able to manage *twenty-eight* inches—one inch more than the famous cloth-yard shafts of our forefathers. (The arrow in the frontispiece is a twenty-nine inch one, pulled twenty-eight inches. As this plate is from a photograph, all the particulars referred to in it may be relied on as correct.) Indeed, it is exceedingly difficult to reconcile the two generally received dogmas of their pull *to* the ear and their *cloth-yard* shafts; inasmuch, as the cloth-yard at that time being but twenty-seven inches, it would appear impossible to get a draw of that length only, so far back as the ear, unless, indeed, we suppose them to have been very short-armed men, or to have kept their bow-arm very much bent whilst shooting. It is more probable that "*to* the ear" meant in a direction *towards* the ear. However this may be, for modern target shooting neither one nor the other is recommended.

To draw to the breast, as many do, is a bad method—indeed, about the very worst, as it circumscribes the pull, most materially diminishes the Archer's power over the bow, and causes the line of sight to be so much above the arrow, that the difficulty of getting an aim (as regards elevation) at the short distances is very great indeed. Moreover, when shot in this way, the arrow flies as if sent from a broomstick rather than from a bow; for the shooter's position is so cramped, so huddled together as it were, that he not only loses a great part of his natural power over the bow, but also that thorough command over the string required for a good loose.

Finally, upon this point of drawing, it should be remarked, that the pull from end to end should be invariably even, quiet, and steady, without jerk or sudden movement of any kind, Some Archers find it extremely difficult to use a bow for any length of time without chrysaling it, and this arises wholly and solely from want of proper attention to this rule. In addition to this, a sudden jerk, especially towards the end, is very likely to pull the arrow and string out of their proper line, and thus spoil the success of the shot.

Chapter XI.

On Aiming.

Prevailing Ignorance on this Point—Absence of Scientific Instruction upon it in all Existing Works—Curious Expedients Resorted to—Their objections—Directions for its Full and Proper Attainment, and its Theory clearly Elucidated—The Point of Aim—A Curious Example—Aiming at Lengths Beyond the Target Distances—Shutting One Eye.

The following observations, be it understood, are intended to apply to those distances only which are fairly within the cast of the bows in use at the present day, and at which accuracy in hitting can reasonably be expected. Beyond 120 or 130 yards, the necessary but excessive arch of the arrow, the unavoidable concealment of the target by the required elevation of the left hand and arm, and the vastly increased effect of wind and weather, all conspire to render hitting the mark a matter much more of chance and guess-work than of skill and scientific practice. Not but what in any case the good Archer will always be superior as against the bad one, even in chance shooting, as he still possesses the advantages that superior judgment and knowledge of his weapon will be sure to give him, and which, to a certain extent, will enable him to control the adverse influences that militate against the correctness and accuracy of his shooting; but the more chance enters into the elements of success, and the more the efforts of skill are baffled by matters out of the power of science to control, the less satisfactory will the pursuit become; and that this is the generally received opinion, as regards distance shooting, amongst Archers of the present day, may be inferred from the fact, that all the numerous Archery societies now existing in the kingdom, with the exception of two or three, limit their distances to 100 yards, as does the Grand National Archery Society itself. I shall now proceed to the discussion of my immediate subject, namely, the method of aiming at what may be called the target distances.

The "aim" is undoubtedly the most abstruse and scientific point connected with the practice of Archery; the most difficult to teach, yet the most neccessary to be taught; upon which all successful practice depends, yet respecting which the most sublime ignorance generally prevails; the want of a due understanding of which is all but universal, yet without which understanding an impassable barrier is presented to the progressing a single step beyond the commonest mediocrity. Ignorance of this fundamental principle it is that causes so many Archers endowed with every quality required to make great and accurate shots—health, strength, correctness of eye, &c.—to stand still, as it were, at a certain point, immoveable, and, if I may coin a word, unimproveable—year after year hammering away in the despairing pursuit of bulls' eyes, without any perceptible improvement or increase of skill, until at last, as I have known in some instances, the whole matter has been given up in sheer hopelessness and disgust at continued ill-success. As if to add to the difficulty of obtaining the command of this most necessary principle of aiming, many of the authors that have treated on Archery have (to judge from their silence) appeared to think such a principle unneeded; whilst others who have noticed it have combined to lead the unfortunate aspirant in the wrong direction. In vain will he search the standard works on the subject through, to find a common-sense or scientific principle laid down to assist him on this point. Ascham will tell him that, "to look at the shaft-head at the loose is the greatest help in keeping a length that can be, but that it hinders excellent shooting;" and afterwards, that, "to have the eye *always* on the mark is the only way to shoot straight." Now, as keeping a length and shooting straight at the same time, are just the two things necessary to hit the mark, and the best *modus operandi* for the one is, according to Ascham, the worst for the other, I do not think much practical benefit is here to be derived from him. He does not even hint at any principle that might by possibility combine the two requisites, neither does he give his readers any further practical assistance, as to getting the straight line and elevation, than is contained in the two above quotations. If he then turn to Roberts, (who, by-the-by, on practical points mostly confines himself to quoting Ascham) he will find from what can be gathered from his (Roberts's) own observations, that the eye is conceived to be an organ of such wonderful power as to be able to accomplish all the Archer may require in the way of elevation, &c. And he is indeed right so far; but whilst he asserts that as the eye is taught, so it will continue to exercise its functions, he totally omits to say how or in what manner it is to be trained so as to arrive at the required powers and capabilities. The author of "The Modern Book of Archery" asserts "that the best, and indeed the only, expedient for attaining perfection in shooting straight, is to shoot in the evening at lights;" and herein, indeed, as so many others have done, he but follows in the wake of Roger Ascham, but improves upon it in favour of the "town resident," by substituting the street gas-lamp opposite his sitting-room, for the paper lantern—policeman A 1, it is presumed, officiating as marker. Waring confines his instruction simply to observing that, "when taking aim, the arrow is to be brought up towards the ear, not to the eye, as many suppose," and that "the Archer must not look along the arrow, but direct at the mark;" and that "the mark is to be visible a little to the left of the knuckles." There are other and smaller works, but they are either plagiarisms from those already named, or ignore instruction upon this part of shooting altogether.

Now, just let my readers imagine such desultory instruction

upon aiming as the specimens quoted, given as regards rifle-shooting, for instance, and it will be instantly perceived how wanting it would be, and how utterly insufficient to enable a man to arrive at anything like excellence in this pursuit. Imagine a man, desirous of hitting a mark at 100 or 200 yards with this weapon, being told to keep his knuckles to the right of the bull's eye, or to keep his eye on it, and trust to his hand following it so accurately as to make his shot all right in the end, without further assistance of any kind. How different are the means actually employed to obtain accuracy here! to what a nicety is the sight regulated! How beautifully calculated to a hair's breadth. A small telescope is even sometimes fixed upon the barrel to insure greater certainty of aim; and even with all these concomitants, the rest is needed by many to make assurance doubly sure. No hand and eye, or lantern and gas-light theories are considered sufficient here. Yet for the bow—an infinitely more difficult weapon to shoot with—such things are gravely set forth as all that are needed, or, at any rate, all that there are to work upon. Hand and eye will do a good deal, no doubt—it will enable a man to throw a stone or bowl a cricket-ball a short distance with tolerable accuracy, or to bring down a partridge or pheasant with a projectile that spreads and covers a space of perhaps two feet in diameter. These and other things of the like nature it may do; but it is comparatively useless when depended on as the only means to enable the Archer to strike with anything like certainty, and with a projectile analogous to a small bullet, a mark much beyond his own nose. The powers of hand and eye are, as with the rifle shot, too limited for him. The truth of this observation may be corroborated by the fact that so many curious devices have been originated by different Archers to obtain some surer means of acquiring certainty. Some will endeavour to find some object to the right or left, above or below the target, which they can apparently cover with the arrow, and which shall yet be about the spot to aim at, so as to cause the shaft to drop into the mark. One I knew of, for sixty yards shooting, used actually to fix a bit of stick into the ground for that purpose. A nice sort of system this to depend upon, on strange grounds and in matches, where no well-placed tree or happily-located stick may happen to be at hand just in the right place. Some have covered the glove of the bow hand with a series of lines of different colours, carrying the eye along one or the other, according as their notion of the line or elevation required, whilst others have improved upon this plan, by making a pincushion of their left hand, by inserting a number of pins in a piece of leather fastened thereon for the purpose, each individual pin serving as a guide for the particular line or elevation wanted at the time. Others, again, have contented themselves with making their left hand their guide, varying its position in conjunction with the mark according to circumstances, high or low, to the right or left hand, as the case might be.

Now these things, and all others like them, are "dodges"—or, as Ascham would call them, shifts—and will never lead to a successful result, or to certainty and accuracy of practice; they may, perhaps, occasionally prove of assistance to those who have no more scientific knowledge of shooting in quiet private practice, where the mind is unexcited and undisturbed, and no distracting influences are likely to arise; but woe to the Archer (in a target-hitting sense) who depends upon them on strange grounds or in matches, or upon any occasion where he may be more than usually desirous of shooting well, for fail him at his need they infallibly will. Strange it is that any such shifts and inventions should ever have been found necessary. Let a gun for the first time be put into a man's hands, and tell him to aim with it, and up it goes at once under the eye, and intuitively he looks at his mark and takes his sight along the barrel. Now the arrow represents precisely an analagous object to this latter; there it is, ready at hand, straight and true, and like as the rifle bullet flies accurately in the direction in which the barrel is held at the moment of discharge, so the arrow will equally, and with the same correctness, fly in the line in which its length lies, and in the direction indicated by itself, when drawn up for the loose—taking for granted, of course, that the shot in all other respects be correctly delivered, that the arrow be a good one, and that no counteracting influence of side wind interfere. The object then to be attained is such a mode of aiming as shall enable the Archer not only to keep his eye upon the point of aim (for this is absolutely necessary for all successful shooting, whether with the gun or with the bow), but at the same time to have a sufficient vision of his mark, and of the length as well as the point of the arrow.

The cause of the great difficulty experienced by the generality of Archers in attaining a satisfactory system of aiming, and the consequent singular devices in vogue for that purpose already mentioned, have appeared to me to arise from a too rigid and mistaken adherence to the supposed old English style of shooting—"pulling to the ear." This *may* have been the method adopted by our forefathers, in the days when great strength and force of shooting was the one thing most sought after, as this method enables the Archer undoubtedly to pull a longer arrow; and thus, the string having a longer distance to act upon the shaft, a quicker and stronger flight is obtained thereby; but I question very much if by this means greater actual power is obtained, but only that the same amount of power is applied in a different manner. This prolonged action of the string, then, upon the arrow, is the whole and sole advantage gained by pulling to the ear; but, in carrying out this method, all scientific principles of aiming must at once be cast aside, because it is impossible when the arrow is once drawn past, and consequently on one side of the eyes, that its true direction can be any longer accurately seen; since, pulled in this way, when to the eye it appears to be pointing to the mark, it is in reality held in a direction far away to the left of it. Hence the reason why some who have written upon the subject of the aim in Archery (assuming at once that pulling to the ear can be the only correct method) direct the learner to keep his bow-hand to the *right* of the mark; and so many Archers aim with one or other of their knuckles, or a particular pin out of their pincushion! I fear I shall be at once anathematised as a heretic for daring to impugn the dear old dogmatic legend of the "pull to the ear;" but I must nevertheless maintain that, with the exception of the advantage above-named, it possesses no recommendation; and if Robin Hood himself adopted this method, and trusted to his hand and eye only, or dodged about with knuckles or pins to obtain an aim, I for one cannot bring myself to believe in his skill, whatever the force of his shot may have been—it may be safely depended on that very few willow wands are to be split in this way. Imagine a man being expected to hit accurately with a rifle with a trigger at his ear, and his eye looking sideways at the barrel: its absurdity at once becomes evident. Yet this is exactly a similar case. I will now, however, proceed to demonstrate what appears to me to be the only true and scientific mode of aiming, and for this purpose it will be necessary, in the first place, to say a few words on those laws of optics which apply to the point in question.

When both eyes are directed to any single object, say the gold of the target, their axes meet at it, and all other parts of the eyes, having perfect correspondence as regards that object, give the sensation of direct vision; but images at the same time are formed of other objects nearer or farther to the right or the left, as the case may be, which may be called the indirect vision; and any object embraced by this indirect vision will be seen more or less distinctly, according to its remoteness or otherwise from either of the axes in any part of their length; and it will be, or at any rate naturally should be, clearest to the indirect vision of that eye to the axis of which it most approximates.

Now, in aiming with the bow, to arrive at anything like certainty, it is necessary to obtain a view of three things, namely, the mark to be hit (which is the gold of the target), the arrow in its whole line and length, (otherwise its *real* course cannot be appreciated), and the point of aim.

It may, perhaps, be as well to explain here, that by the point of aim is meant the spot apparently covered by the point of the arrow. This, with the bow, is never identical with the gold, excepting at one particular distance to each individual Archer, because the arrow has no adjusting sights to make it always so, as is the case with the rifle. As an example, let us suppose an Archer shooting in a side wind, say at eighty yards, and that this distance is, to him, that particular one where, in calm weather, the point of his arrow and the gold are identical. It is clear if he *now* makes them so, the effect of the wind will carry his arrow to the right or the left, according to the side from which it blows. He is, therefore, obliged to aim to one side of his mark, and the point of his arrow, consequently, covers a spot other than that of the gold. And this spot, in this instance, would be to him his point of aim. Under the parallel circumstances of a long range and a side wind, the rifle even would be subject to the same rule.

Now I shall be understood when I repeat, that it is necessary for the Archer to embrace within his vision the gold, the point of the aim, and the true line in which the arrow is directed.

Direct vision, however, can only be applied to one object at a time, and as that object must never in any case be the arrow, I will first proceed to show in what way this must be held, in order to enable the Archer, by means of his *indirect* vision, clearly to appreciate the true line in which it points at the time of aiming, leaving for after discussion the question as to whether the gold or the point of aim should be *directly* looked at.

PLATE VI.

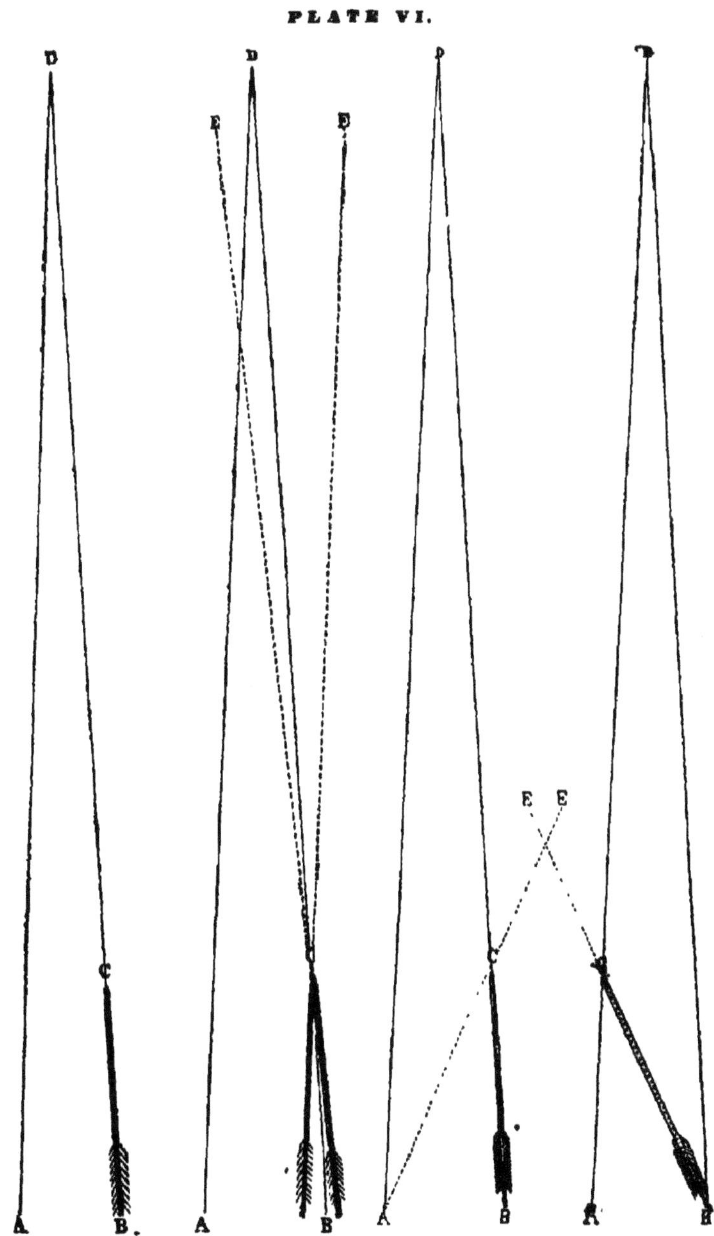

A.B. The two eyes. B. The aiming eye.
C. The arrow.
D The object *directly* looked at.
A.D. and B.D. The axes of the eyes.

A B. The two eyes A. The aiming eye.
C. The arrow.
D. The object *directly* looked at.
A.D. and B.D. The axes of the eyes.

Now it is at once asserted, as an incontrovertible axiom in Archery, that this true line can never be correctly appreciated by the shooter, excepting when the arrow lies *in its whole length directly beneath the axis of the aiming eye.* (The indirect vision of *both* eyes can never be used here, as, if it were, according to the law of optics, *two* arrows would he seen; but this is never the case with the habitual shooter, though both eyes be open, habit and the wonderful adapting power of the eye preventing such an untoward effect equally as well as if the second eye were closed—which, indeed, with many Archers is the case.)

I have said, then, that the arrow, *in its whole length,* must be directly beneath the axis of the aiming eye (see diagram 1, plate 6,) (which I shall here assume to be the *right* one, as in ninety-nine instances out of one hundred is the case,) and it must do so, because otherwise, the shooter will be deceived as to its true line; for so long as the point intersects the axis of the aiming eye, the arrow will appear to that eye to be pointing in a staight line with the object looked at, though in reality directed far away to the right or the left of it. (See diagram 2, plate 6, where the arrow C, though held in the directions C. E., appear to the shooter to be aimed at the object D.)

For instance, suppose the Archer to be shooting at that distance where his point of aim is identical with the gold. He of course brings the point of his arrow to bear upon it, the same as the rifleman would his sights; that is, *the point* intersects the axis of the aiming eye, but if the *arrow itself* be inclined, say to the right of the axis (as in pulling to the ear it would be), it will fly away far to the left of the object looked at—and the converse of this is true also, for if it incline to the left of the axis, it will then fly off to the right. (See diagram 2, plate 6.)

I will produce an example within my own personal knowledge—

a curious, though perfect illustration of all that has been said—and I do so, as it is possible that cases of the like nature may exist, and therefore the description of this one and of its solution may be useful.

An Archer had shot for many years, but always found that if ever his arrow pointed *(to him)* in a straight line with the gold, it invariably flew off far to the left of it—*five or six yards* even at the short distances—(vide diagram 4, plate 6, where the arrow C, though pointing in the direction B. E., appeared to the shooter to be aimed at the object D.) he was therefore obliged to make this allowance and point his arrow *(as it appeared to him)* that number of yards to the right (vide diagram 3, plate 6, where the arrow C, though pointing straight to the object D, appeared to the shooter to be pointing in the direction A. E). During several years he had in vain sought a solution of this anomaly; all could tell him there was something faulty somewhere, but, as everything in his style and mode of action appeared correct, what that something was remained a mystery. Becoming acquainted with him some short time back, he applied to me to solve the riddle, but as I found that the arrow was held perfectly as it should be—directly beneath the axis of the right eye—and that the other important points of the Archer were correct also, I was for a time as much puzzled as any one else could have been. To cut a long story short, suffice it to say, that I ultimately discovered, that though the arrow was held close to, and directly beneath, the axis of the *right* eye, (this being open too,) this Archer actually used his *left* eye to aim with. If the previous observations be considered, it will now at once be seen why the discrepancy between his aim and the flight of his arrow existed; the fact being, that the arrow did not appear to the shooter to be pointing straight till the point intersected the axis of his *left* eye, and consequently until its course was in reality in a direction far away to the left of the mark. (See diagram 4, plate 6.) On closing the left eye, the line of flight and the aim became at once identical, because the eye, under whose axis the arrow was held, became the one with which the aim was taken.

The diagrams illustrating the foregoing observations do not profess to be drawn to scale, but are simply intended to illustrate their principle.

Now, as to whether the *direct* vision should be applied to the mark or the point of aim, the argument is all in favour of the latter. For the point of aim must, necessarily, be in relation to the mark, either in a perpendicular line with it or outside that line: if outside, then the direct vision must certainly be upon the point of aim, otherwise the arrow cannot be directly beneath the line of the axis of the eye, which has already been shown to be necessary; therefore, the only remaining question to be decided is, when the point of aim falls in a perpendicular line with the mark, which of the two should be directly looked at? Here again an argument can be adduced to determine the choice in favour of the former; for when the point of aim is *above* the mark, the latter will be concealed from the right, or aiming eye, by the necessary raising of the bow-hand (as may be proved by the experiment of shutting the left eye); therefore, the direct vision cannot be here applied to the mark, though it may be to the point of aim. There remains then but one other case, namely, when the point of aim falls in the perpendicular line *below* the mark; and here (though either of them may in this case be viewed with the direct vision), as no reasoning or argument can be put forward for violating the rule shown to be necessary in the other cases, and as it is easier to view the point of aim directly, and the mark indirectly, than the contrary, and as uniformity of practice is highly desirable, I strongly recommend that in all cases the direct vision be upon the point of aim. This is contrary to the usual received opinion, which is that the eye should always be intently fixed upon the mark to be hit; but I am very much inclined to think that even those Archers that imagine they do so, will find, as I have done, upon careful experiment, that the point of aim is directly looked at, and not the mark, this being only seen indirectly, except as before stated, when the aim is point-blank; and this is exactly analagous to that part of rifle-shooting where allowance must be made for a strong side wind, at a long range.

My readers must bear in mind that all these remarks, as before stated, are intended to apply only to the target distances, or any lengths within them.

As regards aiming at lengths much beyond these distances, since the mark and the point of aim are too far apart to be sufficiently seen in conjunction, I do not see that any scientific principle can here be laid down for the guidance of the Archer. Practice alone will give him a knowledge of the power of his bow, and the angle of elevation required to throw the arrow up to the mark. If the distance to be shot be a known and fixed one, for instance, two hundred yards, the calculation is more or less attainable; but the great distance renders the aim so uncertain as to prevent anything approaching to the accuracy attainable at the targets. If the mark be a varying and uncertain one, as in roving, the Archer is entirely dependent upon his judgment of distances. This sort of shooting, though very interesting, must be attended with a great amount of uncertainty; but, as in every other case, the more the practice, the greater will be the success.

No rules can be laid down for fixing *where* the point of aim ought to be at any distance, as this is dependent upon so great a variety of circumstances—the strength of the bow, a sharp or dull loose, heavy or light arrows, and the varying force of different winds. This is a matter entirely for the judgment of each individual Archer, and can only be decided by his own practical experience. Indeed, as different winds have such different effects upon the flight of the shaft, it is not until the Archer has arrived upon the field, and actually shot one or two arrows, that even he can be in a position to judge his point of aim for himself. Here the words of Ascham may, with propriety, be quoted:—"The best property of a good shooter is to know the nature of the winds, with him and against him, that thereby he may shoot near to his mark."

Some few Archers are in the habit of shutting one eye when aiming. Now, as it would be anything but interesting or instructive to enter into the discussion of the one and two-eyed theories—a *vexatio quæstio* for centuries, and about which volumes might, and I believe have been written, and no one a whit the wiser in consequence—I shall confine myself to the remarks, that in Archery it is objectionable, whenever the point of aim is above the mark; as in this case, without the use of both eyes, the latter is concealed from the sight altogether by the bow-hand and arm—an instant's experiment will prove the truth of this. Another reason against it is, that though apparently seeming to concentrate the aim, it nevertheless contracts the vision, and, moreover, distorts the face, and interferes with that gracefulness which ought to be one of the attributes of Archery. The fair sex especially will, therefore, be cautious before they adopt so inelegant a system of aiming. There may perhaps be cases, however, where it is almost unavoidable—witness that one of which a detail was given some page or two back. In such instances there is apparently no help for it.

Finally, upon this point of aiming, it should be remarked that, as from the position necessarily assumed by the Archer in shooting, the right eye is the one that comes naturally nearest to the arrow; it is beneath the axis of this one that the shaft should lie. It would hardly have been considered necessary to mention so very obvious a matter, had not a few Archers contracted a habit of actually putting the arrow to the *left* side of their nose, and so under the left eye. Now, as this can serve no useful purpose, has a very awkward appearance, and materially increases the difficulty of keeping the string, when loosed, from striking the left arm (before demonstrated to be fatal to success), the sooner such a habit is got rid of the better. In the exceptional case of the left-handed shooter, of course the contrary of the above is to be followed.

Chapter XII.

Of Holding and Loosing.

The Importance of Holding—Loosing, the last of Ascham's Five Points—Necessity of its Perfect Command—What is, and what is not, a Good Loose—Its Effects upon the Flight of the Arrow—Directions for its Proper Attainment.

HOLDING.

By "holding" is meant keeping the arrow *fully* drawn before it is loosed. Ascham has made this his fourth "point" of Archery, and little or nothing is to be added to what he has said on the subject. "*Holding*," says he, "*must not be long, for it puts a bow in danger of breaking, and also spoils the shot; it must occupy so little time, that it may be better perceived in the mind, when it is done, than seen with the eye when doing.*" This is an entire and exact description of what holding should be, and I shall, therefore, only add that this almost imperceptible pause before the act of loosing serves to steady the arm and correct the aim, and is a grand assistant to the obtaining of a certain and even loose. It is, therefore, with the other points of Archery, most necessary to be cultivated if successful hitting is to be the result.

OF LOOSING.

After the bow is drawn up to its proper extent, and the aim correctly taken, there still remains one more "point" for the Archer to achieve successfully before he can insure the correct and desired flight of his arrow to its mark; and this is the *point of loosing*, which term is applied to the act of quitting or freeing the string from the fingers of the right hand, which retain it. It is the last of Ascham's celebrated "Quintette," and the crowning difficulty the Archer has to overcome, in order to complete the perfection of his shot. Though the last point to be considered, it is not one whit the less important on that account; for, however correct and perfect all the rest of the Archer's performance may be, the result will infallibly prove a failure, and end in disappointment, should this said point of loosing not be also successfully mastered. Upon this, it has been before observed, when treating of the bow, the flight of the arrow mainly depends; and to how great an extent this may be effected by it, may be gathered from the fact, that the same bow, with a like weight of arrow and length of pull, will cast forty or fifty yards further in the hands of one man than it will in those of another, owing solely and entirely to the different manner in which the string shall be quitted; consequently, in target shooting, the aim which may be perfectly correct for one shooter, may be either too high or too low for another, who frees the string in a different manner.

From this it may be gathered how delicate an operation in Archery it is to loose well. To accomplish with evenness, smoothness, and unvarying similarity, it is perhaps the most difficult one of all, and yet for accurate hitting fully as necessary to be attained with all these requisites as any other point of Archery. I think a great misapprehension exists amongst Archers as to what is and what is not a good loose; it being generally thought that if an extreme sharpness of flight be communicated to the arrow, it is conclusive evidence as to the goodness. How often do we hear the observation "What a beautiful loose he has!" though the Archer to whom this remark is applied may be missing arrow after arrow, and vainly endeavouring to hit his mark twice in succession; this encomium being passed upon him merely because his arrow flies keen and sharp. Now, without in the least undervaluing this very excellent quality in the flight of an arrow, and, *so far as it goes*, the goodness of the loose which produces it, I must still maintain that it is not the only requisite; and that unless a *certainty*, as well as a keenness of flight be also obtained, the Archer's "beautiful loose" will be of little avail to him. Undoubtedly the best and most perfect quit of the string would be that which combines both of these qualities; but if the two cannot be obtained together, a slower flight and certainty rise immeasurably superior to the rapid flight and uncertainty: (of course it is meant as regards target, not distance, shooting).

The question then resolves itself into this practical form:—"Is it possible for the same mode of loosing to give extreme rapidity of flight, and, at the same time, certainty of line and elevation?" So far as my experience goes, the answer is decidedly in the negative—not that it is meant to say that a *few* successive arrows may not be accurately shot in this way, but that for any length of time, the uncertainty of flight is sure to be such as to render the average shooting inferior. This difficulty, amounting almost to an impossibility, of obtaining a loose which shall combine great sharpness and certainty of flight at the same time, arises from the fact that such a loose requires (to obtain that sharpness) that the fingers of the right hand be snatched from the string with such suddenness and rapidity as to compromise the second quality of certainty—such a sudden jerk of the string endangering the steadiness of the left arm at the final moment, and, by its unavoidable irregularity, not only having a tendency to drag the string, and, consequently, the arrow out of its proper and original line of flight, but also constantly to vary its elevation. Excepting for distance shooting, then, a *very* sharp loose is not to be recommended; nevertheless, in case he should be engaged therein, the *perfect* Archer should have it under his command.

It must not be supposed, from what has been said, that the exact opposite of the very sharp loose is advocated—that is to say, that the string should be allowed to slip, or loose itself, as it were, without any assistance whatever from the Archer. On the contrary, this mode of quitting the string is the very worst that can be adopted, and one that does more to stay and unsteady the flight of the arrow than any other; in fact, no cast at all can be got out of a bow in this way. But there is a medium between the two extremes, and, leaning rather towards that of sharpness, which, in its practical results, I have invariably found to answer the best. The *modus operandi*, like so many other things connected with Archery, is extremely difficult to describe, if not altogether impossible; but the great characteristic with regard to it is, that *the fingers do not go forward one hair's breadth with the string*, but that their action be as it were, *a continuance of the draw* rather than an independent movement, yet accompanied with just sufficient additional muscular action in a direction away from the bow, and simultaneous expansion of the fingers at the final instant of quitting the string, as to admit of its instantaneous freedom from all and each of them, at the same identical moment of time; for should the string but leave one finger the minutest moment before its fellow, or all or any of them follow forward with it in the slightest degree, the loose will be bad, and the shot in all probability a failure. So slight, however, is this muscular movement, that, though a distinct and appreciable fact to the mind of the shooter, it is hardly, if at all, perceptible to the looker-on, yet, though apparently of so slight a character, so important is it, that the goodness of the loose, and the consequent accurate flight of the arrow, mainly depend upon it. I am painfully conscious of having most signally failed in describing this peculiar mode of loosing in such a way as to enable the learner to understand and practise it; but Ascham's observation, that it is "less hard to be followed in shooting, than to be described in teaching," though not altogether the fact, is not very far removed from it. Had he contented himself with stating that the one was equally difficult with the other, all who have endeavoured to attain it might have agreed with him.

Some Archers use two fingers in drawing, but by far the larger part use three, on account of the greater power the latter mode gives. Provided, however, sufficient strength can be obtained with the first-named method, it may be well recommended, as the string, when quitting the fingers has less surface to glide over, and the accomplishment of the loose is therefore easier; but very few, indeed, can manage a bow of any power without the third finger; consequently the majority of Archers use it.

The position the string should occupy across the fingers is above their first joints, but not too near their tips. On the one hand a too great grip of the string necessitates a drag or jerk (already demonstrated to be unadvisable) to free the fingers, besides giving the string more surface to glide over than is conducive to a smooth and even loose; on the other hand, an insufficient grip of the string deprives the shooter of his necessary command over it, and renders the giving way of the fingers of constant occurrence. Here again, as in so many other instances, the medium between the two extremes is the best, and, it is, therefore, recommended that the string be placed midway between the tips and first joints of the first and third fingers, and rather more towards the end of the middle one—this latter difference being rendered necessary by its greater natural length.

Now, as it is most important, in order to render the loose every time similar, that the string should always occupy exactly the same position on every finger, it has already been advised—when treating of the shooting glove—that stops or guards, to indicate the exact spot on which the string should be placed, should be fastened thereon; and this is especially necessary on the middle finger, which from its greater length, has a tendency when loosing to extend itself, and thus projecting beyond the other two, to present a second obstacle to the escape of the string, after the latter has freed itself from the first and third fingers. Thus the string is sometimes caught upon it, and when this occurs it is unevenly loosed, and causes the shot to be, nine times out of ten, a dead failure.

Especial care must be taken that, whilst loosing, the left arm maintains its position firmly and unwaveringly, and does not give way at the final moment in the slightest degree in a direction towards the right hand, as in this case the arrow is sure to drop short of the mark. It will have precisely the same injurious effect upon its flight as would allowing the fingers of the right hand to go forward with the string. This yielding of the left arm is of more common occurrence amongst Archers than is generally supposed, and is the cause of many an arrow, otherwise correctly shot, missing its mark. All must be firm to the last, and the attention of the shooter never be relaxed for a single instant until the arrow has actually left the bow.

Some Archers have an ugly habit of throwing the left arm and bow, as it were, after the arrow, the instant it has started, as if to lend it a helping hand on its course. Others, again, seem to have a notion that a kick up of the right leg will materially assist its flight. Now these antics, and all others of the like nature, are bad, of course useless, and enemies to all grace and elegance, and, therefore, should be studiously avoided. The shooter should remain perfectly quiescent, *in "statue" quo*—if I may be allowed so questionable a pun—until he is assured of the final destination of his shaft, and satisfied of its success or failure.

Chapter XIII.

On Distance Shooting.

The Divisions—Clout-shooting—Absurdity of the Modern System as a test of skill or strength—The Edinburgh Match—Roving—An agreeable pursuit—Flight-shooting—Length of probable range considered.

Under this head I shall proceed briefly to notice the different kinds of shooting at distances beyond what may be called the regulation target lengths, and, in modern times, they may be classed under three heads—namely, Clout, Roving, and Flight-shooting.

Clout-shooting is so called from the mark being a small white one, termed a "Clout," instead of the ordinary regulation target. The distance at which it is placed varies from nine to twelve score yards: 180 and 200 yards are, I believe, the actual lengths usually shot This kind of shooting, however, is very little practised now-a-days, but two or three societies, out of the many existing in the kingdom, advocating such distant marks; and these, I cannot but think, do so rather from a blind adherence to "tradition," than because of any amusement afforded by such Archery. In former times, when the bow was *the* weapon of war, great force of shooting was the grand desideratum—precision being less required than penetrating power. Hence the laws that were passed at different times, regulating the distances to be shot and the weights of arrows to be used, were made solely for the purpose of keeping up this force of shooting. One of Henry VIII, for example, forbids any but a sheaf or war-arrow being used at eleven score yards or under. In modern days, however, the cultivation of the bow being an amusement, and not a national necessity, it has very properly come to pass that skill and accuracy in its use have become the thing sought after, rather than the mere increase of brute force. Hence all existing societies, with the exceptions mentioned, shoot such distances only as are reasonably within the *hitting* range of modern bows. The exceptional societies alluded to, however, still preserve, under totally different circumstances as regards the practice of the bow, its antiquated distances; influenced, I conclude, by the idea of not degenerating from the force and strength of shooting attained by our ancestors: at least this is the only sensible reason I can think of. If this be the motive, however, they should enforce laws respecting the *weight of the arrow*, as well as the distance shot, and then they might at any rate plead, in defence of their system, their desire to encourage strong shooting. As, however, no such laws are in vogue, the result is that, instead of the bow being strengthened, the arrow is lightened, sensible persons very naturally declining to make a labour of a pleasure. Thus strong shooting is *not* encouraged, for a good fifty lb. bow will carry with ease a light flight arrow a distance of 210 or 220 yards, and this weight of bow is under the average weight used at the target distances. Thus this modern clout-shooting, as a test of strength, is a dead failure. Weak and inexperienced indeed must be the Archer who cannot range an arrow considerably beyond its outside distance.

As a test of skill it is simply ridiculous; for, owing to the excessive but required arch of the arrow in its flight to reach the distant mark, the utter impracticability of scientific aiming (from the impossibility of seeing the point of aim and the mark in juxtaposition), and the great and varying force of winds on a light arrow in a long flight, chance enters so much into the elements of success as constantly to baffle the most experienced shot. As if, too, this chance required still further development, the mode of scoring in use is beautifully adapted to increase its amount; since, instead of the Archer's *average* shooting, each end being reckoned, a hit in the clout, or, failing that (which, nineteen times out of twenty, is the case), the nearest shot to it, is alone allowed to count. Thus, supposing A and B to be two Archers, and that A shoots all his *three* arrows within a foot of the clout, whilst B sends *one* an inch nearer, but the other two any amount of yards off, B alone is allowed to score! though, as regards the relative merits of three shots, A is vastly superior. The chance inseparable from this mode of scoring, however, would be overcome by the good shot in a very short time, and his superiority be speedily made manifest, were he opposed to a single antagonist only; but if the shooters be numerous, it is a very different affair, since the probability of a chance arrow is of course proportionably increased according to the number shot. Thus an extreme case might actually occur where our friend A, competing with twelve others, each shooting twelve ends of three arrows each, might actually beat every one of his competitors thirty-five times out of his thirty-six shots, yet never score once, but be at the bottom of the list after all! For though, each end, his three arrows might be within a foot, or less, of the mark, some single arrow out of the thirty-six opposed to him might each time prevent his scoring. This is, of course, an extreme case; but it will serve to exemplify the system. A curious illustration of the truth of these remarks, as regards the amount of chance in modern clout-shooting, occurred at the Grand National Meeting held at Edinburgh in 1850. On that occasion, as the gentlemen of the Royal Body-Guard principally practised the long distances, exceptional prizes were expressly declared for them, in order that their favourite shooting might come into play. Strange to say, however (yet not strange if the above remarks be properly considered), the first prize for the 180 yards, and the same for the 200 yards shooting, were both won by two gentlemen who had hardly, if ever, shot such lengths in their lives, and who were, moreover, inferior (at that day) at the target lengths to several of the Scottish Archers present. This unlooked-for result not a little nonplussed our friends across the border—as well it might.

In order that my readers may judge whether or no these long distances are fairly within the hitting range of the bow, the following is the result of the shooting at Edinburgh alluded to, though the targets were the usual four-feet ones, instead of the conventional twelve-inch clout—

At 180 yards .... 2,268 shots .... 10 hits!
At 200 yards .... 888 shots .... 5 hits!

Let me, however, do this modern clout-shooting justice. It *has* one recommendation, and it is that of being in some sort "a refuge for the destitute," that is to say, for some whose nakedness and poverty, as regards any real knowledge of Archery, would at once be apparent, where they to appear before the public at a hitting distance. For, as the clout may be said never to be hit (so seldom is this the case); as, consequently, all the arrows, good, bad, and indifferent, equally stick in the ground; and as the spectatots very wisely take care to give the clout a very wide berth indeed (forty or fifty yards is not a foot too much for some of these "long rangers")—the observation that would at once detect a miss at the targets is completely hoodwinked here. Thus, the most excruciating muff—the man who cannot hit a four-foot mark a few yards removed from him—may manage to pass muster as an Archer here, aye, and even strut along the stage of his Archery existence with a comfortable idea of his superiority over the poor, weak, benighted short-range man,—"he never shoots such paltry distances as sixty yards:" he knows better—he would be found out if he did. So, after all, clout-shooting has its advantages.

If it be desirable to encourage the strongest shooting, let such distances be shot and such arrows used, as shall in reality constitute a trial of strength and power over the bow. If skill be the desideratum, let such distances be chosen, and such a mode of scoring adopted, as may give it its fair predominance. This modern clout-shooting is just that happy medium that attains neither end.

ROVING.

Concerning roving, or shooting at rovers, a very few words will suffice. This shooting consists in taking stray and accidental marks, usually at long distances, as the objects to be aimed at, instead of fixed and certain ones. In olden times, when the bow was a weapon of war, the practice of roving was peculiarly valuable, as it tended not only to keep up strong and powerful shooting, but also to give a knowledge of distances and a judgment of lengths peculiarly valuable in battle. In the present day it is seldom or never practised, very few localities, indeed, being sufficiently open for it. It is, however, an interesting amusement, the uncertainty of the distance, and the consequent difficulty of accurately judging it, giving an agreeable amount of excitement.

FLIGHT-SHOOTING.

Flight-shooting (so called from *flight* or light arrows only being used in the sport) is practised solely with the view of experimenting as to the extreme casts of different weights and kinds of bows, or to determine the greatest range to which the power and skill of individual Archers can attain. In modern times it may be safely asserted, that very few shooters (owing partly to a want of practice in flight-shooting) can cover a distance of 300 yards; and to attain this range, a bow, and a good one too, of at least sixty-two or sixty-three pounds, must not only be used, but thoroughly mastered, not merely as regards the drawing, but in respect of quickness and sharpness of loose also; for, as before remarked, the rapidity of the arrow's flight depends principally upon this latter qualification. Thus an Archer may be able to *draw* a bow of seventy or eighty lbs. and yet very likely be unable to *loose* properly one of more than fifty-six or sixty lbs., and, consequently, he will not be able to increase his range at all in proportion to the increase of power in his bow. Indeed, he will in all probability shoot further with the weaker bow within the command of his loose than with the much more powerful one beyond it, however much the latter may be under the power of his pull. Now, without hesitation, it is affirmed that there is hardly an Archer living (if there be one) who can loose properly a bow much over seventy lbs.; though there are many that can draw seven-five lbs. or eighty lbs., and a few perhaps some pounds beyond even this; if this be so, a range much over 300 yards is not likely to be attainable.

Some years back Mr. Muir, of Edinburgh, made many experiments with strong and medium power bows, with the view of testing the possibility of accomplishing 300 yards; but, though an Archer of great power and experience, he found that with a bow of from fifty-eight to sixty-two lbs. he could shoot further than with a stronger one, and that with that weight of bow he could not quite reach the desired distance. Afterwards, however, with a Turkish horn bow and flight-arrow, he accomplished a measured range of 306 yards. Mr. Roberts, in his "English Bowmen," (published in 1801), states that Mr. Troward shot repeatedly, up and down, and in the presence of many spectators, a measured length of 340 yards, with a self yew bow of sixty-three lbs.

In this kind of shooting I have, personally, had very little experience, but in the autumn of 1856, in the presence of a brother Archer, I succeeded, upon several occasions, in exceeding the 300 yards, the longest shots being 308 yards, with a slight wind in my favour, and in a perfect calm 307 yards one foot; the ground was carefully measured with the tape. The bow used was a sixty-eight lb. self yew, taken at random from Mr. Buchanan's stock, and was by no means remarkable for quickness of cast, though it has since proved an excellent target bow.

I believe, therefore, that, *with practice*, 300 yards is fairly attainable by many Archers of the present day, and that several might even reach very considerably beyond it; but to attain to this distant shooting, a particular study of itself would be required, as it is a totally different matter from target practice.

However, experiment alone will enable any Archer, curious as to his powers of distant shooting, to determine the length he will be enabled to reach. To that he is, therefore, recommended.

Chapter XIV.

On Ancient and Modern Scoring.

Best Shots of the Toxophilite Society—Mr. Brady—Mr. Crunden—Mr. Palmer—Mr. Cazalet—Mr. Shepheard—Result of Mr. Waring's Arm-striking theory—Mr. Anderson, the Incomparable Shot—Scores of more Modern Archers—First and Second Scores of all the Grand National Meetings.

In the present chapter I propose presenting to my readers a few specimens of ancient and modern scoring. The term "ancient," however, must be considered as used only in a comparative sense; for the earliest period to which I am able to refer goes no further back than 1795, some few years after the first establishment of the Toxophilite Society, and the subsequent revival of Archery. Anterior to this period I have been unable to obtain any authentic records, none such having been kept, or, if kept, having in the lapse of time been misplaced or lost. For the less modern scores about to be given, the reader is indebted to the books of the Toxophilite Society, some of the earlier of which have fortunately remained in existence to the present day, whilst others, including the whole from 1806 to 1834, have been unaccountably lost. As this Society has always, from its first commencement until now, numbered amongst its members some of the best, and generally *the* very best Archers of the day, the specimen scores given may be fairly looked upon as the *good* shooting for the times during which they were achieved; and, consequently, a pretty accurate opinion may be formed as to the capabilities of the then magnates of the bow. The result of their comparison with the shooting of the present day will show what rapid strides in advance Archery has made since the establishment of the Grand National Meetings has held out an adequate inducement for its proper study and practice. The great scores of times anterior to these Meetings now cut but a sorry figure indeed. May we not hope that, some years hence, the same may be said of the great performances of the present time?

The first score I shall give is that which won the Prince's annual bugle in 1795; and it is given more especially because Mr. Waring, in his treatise, calls it "undoubtedly *very great* shooting." The Toxophilite records give it as follows:—

276 shots, 90 hits, 348 score.

Mr. Brady was the shooter, and the distances were 100, 80, and 60 yards, at four, three, and two feet targets respectively (the smaller targets at the shorter distances may be considered, perhaps, a trifle easier to score at than the full sized ones at 100 yards). Thus, according to Mr. Waring, something *worse* than *one* hit out of *three*, at 100 yards, was *very great* shooting in his day. Down to 1805 this score was beaten only once, and that occurred in 1797, for the same prize, and at the same marks and distances, when Mr. Shepheard scored as follows:—

252 shots, 88 hits, 358 score.

There appears to have been a bye-law respecting this annual bugle, preventing the same member from winning it more than one year; so, in 1799, it was gained by Mr. Waring, with 53 hits and 185 score. The number of arrows shot in this instance is not given; but as the shooting lasted two days, it could not well have been less than what appears to have been the ordinary round for this prize—namely, 252 arrows, equally divided between the three distances.

By far the best average shot of that era was, indisputably, Mr. Crunden; though several, namely, Messrs. Palmer, Shepheard, and

Cazalet, sometimes surpassed him, and generally were, one or other of them, close at hand. The first-named gentleman, indeed (Mr. Palmer), made the greatest single day's shooting during the ten years to which I am able to refer, having scored as follows at 100 yards :—

192 shots, 91 hits, 379 score.

Subjoined are the three best scores of Mr. Crunden during the same period (taking the hits as the criterion), the distance being still 100 yards :—

192 shots,	85 hits,	331 score.
192 ,,	79 ,,	419 ,,
192 ,,	76 ,,	238 ,,

The like of Mr. Palmer :—

192 shots,	91 hits,	379 score.
204 ,,	78 ,,	268 ,,
192 ,,	57 ,,	205 ,,

The like of Mr. Shepheard :—

192 shots,	71 hits,	253 score.
192 ,,	61 ,,	233 ,,
192 ,,	57 ,,	211 ,,

The like of Mr. Cazalet :—

192 shots,	77 hits,	245 score.
240 ,,	73 ,,	271 ,,
192 ,,	62 ,,	184 ,,

The amount of benefit to be derived from Mr. Waring's "arm-striking" theory may be estimated by reference to this gentleman's three best scores :—

192 shots,	34 hits,	130 score.
192 ,,	41 ,,	153 ,,
204 ,,	41 ,,	157 ,,

I am happy in being able to furnish my readers with two of Mr. Anderson's scores—that "incomparable" *(sic)* Archer, according to Mr. Roberts; they are as follows, the distance being 100 yards:—

192 shots, 37 hits, 137 score.
216 „ 46 „ 182 „

If these are fair average specimens of this gentleman's shooting, his "incomparability" is in the very opposite direction to that intended by Mr. Roberts.

It must be remarked, concerning the foregoing scores, that they are taken from those records only where the number of arrows shot is stated; this, though only occasionally the case, is sufficiently often as to render the specimens given a good average criterion of the shooting of that day. Scores made at 100 yards are given principally, as at the eighty and sixty yards different sized targets from those at present in use were then shot at.

For want of the records from 1805 to 1834, I am now obliged to jump at once to the latter year. Whether during this interval the shooting improved or not is a question that must be left undecided. Probably not. But, if it did, it must again have retrograded, as from 1834 to 1844 (during which latter year the first National Meeting was held) the shooting appears to have been of about the same average character. Indeed, I am unable to find an instance up to the latter date of Mr. Palmer's best score, already given, having being either beaten or equalled. The nearest approach to it is a score of Mr. Peters's, as follows:—

192 shots, 88 hits, 328 score.

Of this era this gentleman appears to have been the Robin Hood, though Messrs. Norton, Robinson, Arabin, and Smyth, contested the palm with him, and not always without success. The St. George's

Society (likewise a London Club) also possessed several shots equal in skill to these.

Still the days when it was considered impossible to put in *half* the arrows at 100 yards (excepting as a rare feat) were dragging their slow length along. Indeed, I have seen a letter as late as 1845, from good old Mr. Roberts, who was well acquainted with the powers of all the best Archers for the preceding half-century, in which he states "he never knew but one man that could accomplish it." From what has been already stated, it will be seen that no single *recorded* instance of its having been achieved at all is to be found up to that date—at any rate, as far as the Toxophilite books are concerned. It ought, however, here to be mentioned that, up to this date, the scoring part of the target measured only three feet ten inches in diameter, the Archer not being allowed to score the hits in the edge beyond this, which was then called the "petticoat." Also another rule then prevailing militated against the scoring, namely, that when the arrow struck *two* circles the *least* only was marked, whereas at the present time the whole of the target (four feet) scores, as well as the higher numbered circle. Six per cent, however, on the hits, and ten per cent. on the score, will be a most liberal allowance to counterbalance the drawbacks alluded to. In comparing, then, the foregoing scores with those about to be mentioned, the reader must bear in mind the above observations.

In 1844 the first Grand National Meeting was held, and the spur it and the succeeding meetings gave to the pursuit of Archery, and the beneficial effects of a proper inducement to its practice, soon became apparent. Amongst the first to emerge from the "slough of despond," in which the art had so long slumbered, were Messrs. Bramhall, Maitland, Muir, Hutchons, and some others, followed shortly afterwards by Messrs. Moore, Garnett, Ford, Hilton, and other good Archers and true. Before, however, giving any of the scores of these latter magnates, it is necessary to bear in mind the distinction between match-shooting (more especially as regards the National Meetings) and private practice. Many, taking an interest in the subject of Archery, and hearing of the great things done at the present day when compared with the achievements of a former period, immediately refer to the records of these National meetings to find these great scores, and, to their surprise, discover that, excepting in one or two instances, the shooting, when compared with that of forty or fifty years back, is very little better. This, however, is not a just comparison, since, no such gatherings existing in those days, the best scores made then (many of which I have given) were obtained either in the ordinary practice-days of the Club, or at their quiet private matches. To shoot at the Grand National Meeting is a totally different affair, as every Archer who has tried the experiment is too well aware. Here the excitement of the occasion, the number of competitors, and the vastness of the assemblage, are enough to upset the firmest nerves—they need, indeed, be of iron, to remain totally unaffected. Added to which, it is an opportunity of making a good score that occurs but once a year, and even this is often marred by unfavourable weather. Hence every Archer, I may say, I think, without exception, falls below his level at this match; consequently, his real powers, excepting amongst the initiated, cannot be judged thereby. In comparing, therefore, the scores of this day with those of a prior date, such only must be looked at as are shot under similar circumstances.

In the following scores I do not pretend to give specimens of the shooting of *all* the good Archers of the day, but of such only as, through some authentic channel or other, have come to my own *personal* knowledge. They, however, will be sufficient for the purpose I have in view, namely, to show the great development of the powers of the bow that has taken place of late years.

The following are the first and second scores in a match that took place the 9th of October, 1850, on the Toxophilite ground, for a handsome silver cup, presented by W. Peters, Esq., the distance being 100 yards :—

Mr. H. A. Ford,	216 shots,	166 hits,	628 score.
Mr. C. Garnett	„	125 „	521 „

The next score is one with which, on the Toxophilite ground, I won a handicap prize of £15, in June, 1854, thirteen competitors :— 100 yards, 96 shots, 79 hits, 373 score; 80 yards, 72 shots, 71 hits, 325 score; 60 yards, 48 shots, 47 hits, 313 score. Total—

216 shots, 197 hits, 1011 score.

In November, 1851, a friendly passage of arms between Messrs. Ford, Bramhall, and Moore, resulted in the following score—the double York round of 144 arrows at 100 yards, 96 at 80 yards, and 48 at 60 yards, being shot :—

Mr. H. A. Ford,	288 shots,	262 hits,	1414 score.
Mr. Bramhall	„	250 „	1244 „
Mr. Moore	„	223 „	1045 „

The 100-yard part of the shooting was very good; Mr. F. getting at this distance 127 hits, 617 score; Mr. B. 114 hits, 504 score; and Mr. M. 100 hits, 440 score. This is not, however, one of the most favourable specimens of this last-named gentleman's shooting. The following is a better one, obtained in private practice—still the double York round :—

288 shots, 252 hits, 1288 score.

Here is also an excellent double York round of Mr. Bramhall's :—

288 shots, 256 hits, 1322 score.

At 100 yards, 117 hits, 535 score; at 80 yards, 91 hits, 497 score; at 60 yards, 48 hits, 290 score.

Also a goodly specimen of 60-yards shooting by the same gentleman—the St. Leonard round:—

75 shots, 74 hits, 504 score.

The following are two good examples of 100-yard shooting, achieved by Mr. Charles Garnett:—

72 shots, 61 hits, 269 score.
72 „ 58 „ 288 „

One of the most promising shots of his day, both for style and accuracy, was Mr. E. Maitland, of the Queen's-Park Archers. Unfortunately for the cause of Archery he went abroad, and thus his career as a bowman, for a time, came to a conclusion. He has since, however, returned to this country, and we anticipate seeing him speedily resume his position amongst the first Archers of the day. The scores that follow were his best. The St. George's round:—100 yards, 36 arrows, 25 hits, 97 score; 80 yards, 36 arrows, 34 hits, 190 score; 60 yards, 36 arrows, 36 hits, 196 score: total—

108 shots, 95 hits, 483 score.

Also a good St. Leonard's round, 60 yards:—

75 shots, 75 hits, 467 score.

Another member of the same Society, Captain Flood, has also achieved some very creditable shooting, more especially at 60 yards; for instance, for 36 arrows, 36 hits, 222 score; and for 75 arrows, a score of 417.

The St. George's Club have turned out some very excellent Archers, amongst whom may be numbered Messrs. Hutchins, Marr, and Heath. I subjoin two specimens of Mr. Marr's best shooting. The St. George's round:—100 yards, 36 arrows, 24 hits, 114 score; 80 yards, 36 arrows, 32 hits, 118 score; 60 yards, 36 arrows, 35 hits, 181 score: total—

108 shots, 90 hits, 413 score.

Also a better specimen of 60 yards shooting by the same gentleman :—

36 shots, 35 hits, 225 score.

The following is one of Mr. Heath's best scores, the St. George's round :—25 hits, 89 score; 31 hits, 139 score; 35 hits, 203 score; total—

108 shots, 91 hits, 431 score.

The National distances, until of late years, have not been much practised in Scotland; consequently our friends over the border have not as yet achieved similar scores to those here given. At their point-blank distance, however, (100 feet) Mr. Watson, of the Royal Company, has put *nine* consecutive arrows into a *four-inch* paper; and Mr. Muir *five*—two undoubtedly clever performances. The latter gentleman, at 100 yards, has also put in thirty-eight arrows out of forty-eight; several times twenty-four arrows out of twenty-five; and similar achievements. His best score, however, to my mind, is the following, distance between 20 and 30 yards :—

Two shots, two hits, score, a hawk and a crow (fact).

Under the risk of being considered egotistical, but to oblige the request of several correspondents, I now give the three following speciments of my private practice—I need hardly say my best. The first two are the single York round of six dozen, four dozen, and two dozen. At the first I made (with an Italian self yew-bow of Mr. Buchanan's, and 5s. arrows of Mr. Muir's) 71 hits, 335 score, (missing the 59th shot); 48 hits, 272 score; 24 hits, 158 score; giving a total of—

144 shots, 143 hits, 765 score.

At the second (with a yew-backed yew-bow and same arrows) 66 hits, 344 score; 47 hits, 301 score; 24 hits, 164 score; total—

144 shots, 137 hits, 809 score.

The following is a St. Leonard's round, at 60 yards:—28 golds, 37 reds, 7 blues, 3 blacks; total—

75 shots, 75 hits, 555 score.

All these scores were made in the public gardens at Cheltenham, in the presence of many persons. With the private shooting of many excellent Archers, such as Messrs. H. Garnett, Hilton, Mallory, &c., I am unacquainted, and therefore unable to give specimens of it.

As a matter of considerable interest to the general body of Archers, I shall now give the names of the first and second winners (ladies and gentlemen) at all the Grand National Meetings up to the present time—also their gross hits and scores. It must be borne in mind that the number of arrows shot at all these gatherings (with the exception of the first, when only *half* the quantity were shot) was 144 at 100 yards, 96 at 80 yards, and 48 at 60 yards, for the gentlemen; and 96 at 60 yards, and 48 at 50 yards, for the ladies,—excepting in the instances that will be specially referred to.

1844—AT YORK.

1. Mr. Higginson	...	63 hits,	221 score.
2. Mr. Meyrick	...	58 „	218 „

No ladies appeared at this Meeting; and, as already mentioned, the gentlemen only shot one-half the quantity shot since.

1845—AT YORK.

Ladies.—1.	Miss Thelwell	...	48 hits,	186 score.
„ 2.	Miss Townshend	...	45 „	163 „
Gentlemen.—1.	Mr. Muir	...	135 „	537 „
„ 2.	Mr. Jones	...	129 „	499 „

At this Meeting the ladies shot 144 arrows, at 60 yards only.

1846—AT YORK.

Gentlemen.—1. Mr. Hubback	...	117 hits,	519 score.
„ 2. Mr. Meyrick	...	117 „	517 „

Close fighting indeed! No ladies shot at this Meeting.

1847—AT DERBY.

Ladies.—1. Miss E. Wylde	...	65 hits,	245 score.
Gentlemen.—1. Mr. Muir	...	153 „	631 „
„ 2. Mr. Maitland	...	131 „	549 „

The ladies again shot at 60 yards only, the number of arrows being 144, as before.

1848—AT DERBY.

Ladies.—1. Miss J. Barrow	...	47 hits,	167 score.
„ 2. Miss Temple	...	44 „	160 „
Gentlemen.—1. Mr. Maitland	...	135 „	581 „
„ 2. Mr. Bramhall	...	132 „	514 „

During both the days of this match a very strong wind prevailed, accompanied with constant showers. The difficulty of scoring was, consequently, very much increased. At this Meeting the ladies shot 72 arrows at 60 yards; and 72 arrows at 50 yards.

My first appearance at these tournaments—my place at the finish being so low down in the list, that I have never to this day had the moral courage to enquire how far from *the bottom* it was. I mention this in the hope it may prove a consolation to many other young Archers, who having attended their first National Meeting with great hopes of success, founded upon the steadiness and goodness of their private practice, have returned home, as I did, sadly disheartened and crest-fallen,—not because of their failure in getting a prize, but on account of the excessive falling-off in their anticipated scoring. I cannot too often impress upon these the fact, that shooting at the National Meeting is totally different from private practice, or small match-shooting; and rare indeed is it that the Archer attends one of them for the first time without a signal failure to his hopes, and a score very much below what his private shooting had led him to look for.

1849—AT DERBY.

Ladies.—1.	Miss Temple	...	55 hits,	189 score.
„ 2.	Miss Mackay	...	43 „	163 „
Gentlemen.—1.	Mr. P. Moore	...	173 „	747 „
„ 2.	Mr. H. A. Ford	...	177 „	703 „

The weather again unfavourable, a good deal of wind prevailing, and many showers. The Champion's Gold Medal was first awarded at this meeting. This was gained by myself, though obtaining only the second prize,—the medal being given for the greatest number of *points* gained by any Archer. These points are reckoned in the following manner:—*Two* for the gross score, *two* for the gross hits. *One* for best score at 100 yards, and *one* for best hits at ditto; and the same at the 80 and 60 yards. This makes ten points in all. I gained five points, Mr. Moore four points, and Mr. Attwood one point.

1850—AT EDINBURGH.

Ladies.—1.	Mrs. Calvert	...	47 hits,	161 score.
„ 2.	Miss E. Foster	...	42 „	156 „
Gentlemen.—1.	Mr. H. A. Ford	...	193 „	899 „
„ 2.	Mr. C. Garnett	...	166 „	638 „

The weather wet, but little or no wind. The Medal awarded to myself, gaining all the points. The ladies on this occasion shot 72 arrows at 60 yards; and 36 arrows only at 50 yards.

1851—AT LEAMINGTON.

Ladies.—1.	Miss Villers	...	108 hits,	504 score.
„ 2.	Mrs. Thursfield	...	75 „	293 „
Gentlemen.—1.	Mr. H. A. Ford	...	193 „	861 „
„ 2.	Mr. Bramhall	...	178 „	760 „

Compare Miss Villers' score with those of the ladies gaining the first prize that preceded her. What a rapid stride in advance! This lady was the first to demonstrate what the bow could do in the hands of the fair sex, and so deservedly obtained *for herself* the reputation of the first lady Archer in the kingdom, a reputation since amply upheld under her married name of Mrs. Davison. The second gross score amongst the gentlemen was obtained by Mr. Heath, of the St. George's Club. The second prize, however, was awarded to Mr. Bramhall for gross hits, Mr. Heath's hits being only 168, with a score of 776. The Champion's Medal again awarded to myself, gaining nine points, Mr. Heath gaining one point, *viz.* the greatest score at 80 yards.

1852—At Leamington.

Ladies.—1.	Miss Brindley	...	84 hits,	336 score.
„ 2.	Miss M. A. Peel	...	84 „	330 „
Gentlemen.—1.	Mr. H. A. Ford	...	188 „	788 „
„ 2.	Mr. Bramhall	...	184 „	778 „

This match had a most exciting *finale* between the first and second gentlemen winners. When the last *three* arrows alone remained to be shot, Mr. Bramhall was two points in score a-head. It was then a simple question of nerve, and I conclude mine was best, as I scored fourteen to my worthy opponent's two. The Champion's Medal awarded to myself, gaining six points, Mr. Bramhall two points, and Mr. Wilson (of York), two points.

1853—At Leamington.

Ladies.—1.	Mrs. Horniblow	...	89 hits,	365 score.
„ 2.	Miss M. A. Peel	...	84 „	364 „
Gentlemen.—1.	Mr. H. A. Ford	...	202 „	934 „
„ 2.	Mr. Bramhall	...	167 „	733 „

The Challenge Silver Bracer for the ladies was, this year, presented by the West Norfolk Bowmen. This prize is awarded on the same principle as the Champion's Medal—namely, for the greatest number of points. It was gained by Mrs. Horniblow, another star in the Archery hemisphere, this lady gaining six points, and Miss M. Peel two points.

The Champion's Medal awarded to myself, gaining all the points.

1854—At Shrewsbury.

Ladies—1.	Mrs. Davison	...	109 hits,	489 score.
„ 2.	Mrs. Horniblow	...	96 „	398 „
Gentlemen—1.	Mr. H. A. Ford	...	234 „	1074 „
„ 2.	Mr. Bramhall	...	176 „	748 „

The Challenge Bracer awarded to Mrs. Davison (*neé* Villers, and who did not shoot at the previous Meeting), this lady gaining seven points to Mrs. Horniblow's one.

The Champion's Medal to myself, gaining all the points.

1855—at Shrewsbury.

Ladies.—1.	Mrs. Davison	...	115 hits,	491 score.
„ 2.	Mrs. Horniblow	...	103 „	437 „
Gentlemen.—1.	Mr. H. A. Ford	...	179 „	809 „
„ 2.	Mr. Bramhall	...	175 „	709 „

Weather wet and windy. The Challenge Bracer was again awarded to Mrs. Davison, that lady gaining seven points; Miss Clay one point for score at 60 yards.

The Champion's Medal to myself, gaining nine points, the tenth (being hits at 60 yards) being a tie with Mr. Wilson.

1856—At Cheltenham.

This Meeting witnessed a larger gathering of Archers, both ladies and gentlemen, than any previous occasion, 72 of the former and 112 of the latter assembling at the targets. The weather was rather against the shooters, owing to the excessive heat and glare. The result was as follows:—

Ladies.—1.	Mrs. Horniblow	...	109 hits,	487 score.
„ 2.	Mrs. Davison	...	103 „	461 „
Gentlemen—1.	Mr. H. A. Ford	...	213 „	985 „
„ 2.	Mr. Bramhall	...	191 „	785 „

Mrs. Davison, through sudden indisposition, was unable to shoot the last six arrows, otherwise the contest for the ladies' supremacy would have been close indeed. The Challenge Silver Bracer was awarded to Mrs. Horniblow, gaining six points to Mrs. Davison's one point, the hits at 60 yards being a tie. The Champion's Medal came to myself, for the eighth time, gaining all the points but two, these latter falling to the lot of my old opponent, Mr. Bramhall, for score and hits at 100 yards.

1857—At Cheltenham.

Ladies.—1.	Miss H. Chetwynd	...	128 hits,	634 score.
„ 2.	Mrs. Davison	...	114 „	548 „
Gentlemen.—1.	Mr. H. A. Ford	...	245 „	1251 „
„ 2.—	Mr. G. Edwards	...	188 „	786 „

This was again a very successful Meeting, and the scoring showed a great and general improvement upon former years. The score made by Miss H. Chetwynd is quite unparalleled in match shooting, and with few, if any, exceptions, in private practice. Mrs. Davison's shooting was also splendid; as was Mrs. Horniblow's—the latter lady, indeed, exceeding the former by eight hits, whilst she was only eight behind her in score. Nor must I omit to notice the shooting of Mrs. Blaker, who, though appearing at these Meetings for the first time, gained the honourable position of fourth, with the excellent total of 108 hits, 496 score. The advancement of the gentlemen kept pace with that of the ladies; for, whereas the average of the five best scores in former years had never exceeded

179 hits, 757 score, in this year it reached 191 hits, 861 score. In the year 1849, Mr. P. Moore accomplished 747, and, in 1851, Mr. Heath 776. But, with these exceptions, until the present year, Mr. Bramhall and myself alone reached 700. At this Meeting, however, as many as six competitors exceeded that number; Mr. Bramhall, for the first time during the last eight Meetings at which he had contended, failing to secure the second place—ill health and consequent want of practice having seriously diminished his chance of success. It is gratifying to me to be able to state that several of the leading Archers on this occasion attributed their high position in the prize list to their careful following out of the principles and directions laid down in this work. The Silver Bracer was awarded to Miss H. Chetwynd, and the Champion's Medal to myself, each having gained all the points.

1858—At Exeter.

Ladies.—1.	Mrs. Horniblow	...	101 hits,	457 score.
„ 2.	Mrs. St. George	...	94 „	428 „
Gentlemen.—1.	Mr. H. A. Ford	...	214 „	1076 „
„ 2.	Mr. G. Edwards	...	187 „	817 „

There is an apparent falling off in the shooting of this Meeting, as compared with the previous one; but I think it is apparent only, as the roughness of the weather and the difficulty of the ground, as compared with that of Cheltenham, will amply account for the diminution in the average scoring. Mr. Edwards, however, exceeded his score of the previous year, topping 800—a feat never performed before at these gatherings, except by myself. Several new shots, also, both ladies and gentlemen, made a most promising *debût*, more especially, amongst the ladies, Mrs. St. George, who attained the second place; and, amongst the gentlemen, Mr. George, who came out third on the prize list. The Silver Bracer was awarded to Mrs. Horniblow, gaining four points out of the eight—Miss H. Chetwynd, Miss Turner, and Lady Edwardes, obtaining one each, the remaining point being a tie between the Lady Champion and Miss Chetwynd. The Champion's Medal to myself, for the tenth time, gaining all the points.

This concludes the record of the National Meetings up to the present time, and I cannot close this chapter without congratulating my brethren of the bow upon the very evident progress of Archery in public estimation, evidenced, not only by the constantly increasing attendance both of Archers and spectators at those Meetings, but also by the formation of new societies in every part of the kingdom, and the institution of large and influential gatherings, open to all, in localities hitherto unused to them. The Grand National Meeting was established in 1844, and for that and the five following years, the average attendance of lady shooters was six, of gentlemen, 74; whilst during the last six years, the ladies have averaged 56, and the gentlemen 91—a pretty evident proof of the increasing popularity of this most healthy and delightful exercise. In further corroboration of this, I need only refer to the Meetings established at Leamington, in 1854, by Mr. Merridew, and that at Aston Park in the present year, all of which have been attended with the highest success, and have afforded unbounded satisfaction to every Archer who has had the good fortune to be present at them; and long may they flourish to the promotion of good feeling and fellowship between all ranks of society!

Chapter XV.

Robin Hood—Distance and Accuracy of the Shooting of his time—Did he shoot in a Modern Hat?—Social Character of Archery—Concluding Observations to the Young Archer; also to the Old One—A Short Address to the Fair Sex—A Farewell.

It was my intention, upon setting out, to devote a chapter to a brief sketch of the life and feats (more particularly where the latter were connected with Archery) of that renowned outlaw, bold Robin Hood. Having, however, after considerable trouble and research, been unable to discover a single fact or legend, probable or improbable, concerning him and his "merrie men," that had not already been written upon *usque ad nauseam*, I finally concluded it would serve no useful purpose were I to add another to the already long list of publications on the subject. As regards his life generally, works of all sorts abound; every age has been suited, every taste, whether for the truthful or the marvellous, accurately fitted. The subject, in short, has been entirely exhausted. As regards his feats with the bow, all those that rest on a shadow of foundation (and none have any better than some old song or legend, written some centuries after his death) have been thoroughly sifted and brought to the light of reason and common sense in Mr. Roberts's work, so often alluded to. In respect of his distance shooting, for instance, he proves, satisfactorily enough, that Robin could *not* shoot a mile—a feat some legends give both him and Little John the credit of being able to perform! and, considering it would take at least *three* of the strongest men in the world put together to pull and loose a bow of sufficient power to do it, my readers will be very much disposed to agree with him. In respect of the *accuracy* of his shooting, splitting willow wands and nocks of arrows (at any rate, if we are to believe novelists) appears to have been his *average!* That he could do the first tolerably often is likely enough—many Archers of the present day could do the same; but to say that he could do it *to a certainty*, is a trifle beyond even the most vivid imaginative power of belief. As for the nock-splitting, it is only necessary to say that beyond fifteen or twenty yards the nock would not be *visible*. I will not, therefore, insult my readers by arguing the possibility or otherwise of Robin's being able to split it at four or five times that distance *whenever he chose!*

Whilst, however, withholding credence from such impossible fictions as those above alluded to, it may very well be a question, whether Archery had not, in Hood's time, attained a greater development in respect of *skill* than is the case at the present day—regarding its greater *force* there can be no dispute. The bow was then *universally* practised, now it is but partially so; therefore we may very well suppose that, as the number of Archers at that time was so much greater, a proportionately larger number of first-rate shots would be produced, and that out of this latter class, consequently, a greater chance would exist of the issuing of *the* pre-eminent shot of all. But whilst the full benefit of this argument is given to Robin Hood, it must not be forgotten that in his day *force* was the great desideratum, and that very strong shooting, so far as our experience goes, is anything but likely to be the most accurate. Does any Archer of the present day, where the prizes are awarded for accuracy, venture to use a bow fully up to the power of his pull? Has not almost every shooter, on the contrary, ten, twenty, or even more pounds, of actual power in him, than he finds it good policy to use? And he does not use it, because experience has taught him that when his muscles are strained beyond a certain point, uncertainty and inaccuracy are sure to follow. In former days, however, the thing was different—*force* was everything in the shooter, *accuracy* being but a secondary consideration; for, as the long-bow was essentially a weapon of war, it mattered little whether the Archer hit the man he aimed at or the one next him, provided the arrow had sufficient power to penetrate the finely tempered armour then in use—shooting at a body of men, however few they might be, to miss all was an exceeding improbability.

Moreover, it may very well be disputed, whether the weapons themselves had the same excellent workmanship then as now; and upon this great certainty of hitting very much depends. The foregoing pages have been written to very little purpose indeed, if I have failed to shew how the slightest inequality or imperfection in the tackle, be it bow, arrow or string, may mar the perfection of the shot. If then, the bowyers of the present day exceed those of the former period in the excelling of their weapons, it is more than probable that a corresponding improvement results to the accuracy of the shooting. But do they? This cannot be decided, so few specimens of the old weapons remaining in existence at the present day, by which to form a comparison.

Each will probably form his opinion as to the skill attained in the first ages of the English long-bow, as compared with that of the present day, according as his taste lies in the direction of the probable or the marvellous; and there the matter must be left.

There is one thing, however, connected with Robin, that I do feel quite confident about, and that is, that he did *not* wear an ordinary modern black or white hat. He was by far too sensible a fellow, depend upon it, to use a head-covering, at all times sufficiently tight and uncomfortable, but especially so when the wearer is more than usually exposed to heat and sun, as the Archer is. *Fancy this bold outlaw in a white four-and-nine!* Alexander the Great in an umbrella, Socrates in Hessian boots, might be parallel cases!

Besides, Robin knew too well, that anything likely to impede the correct action of the string—as the brim of a modern hat very often does—would effectually prevent the indulgence of his willow-wand and nock-splitting propensities. Therefore, he doubtless avoided anything of the kind. I commend these reflections to one or two of the societies of the present day.

Robin must have been a social fellow—all accounts of him prove this. Indeed it is one of the great advantages of this noble pastime of Archery, that it induces this very quality. There is nothing more productive of kindly feeling and hospitality in a neighbourhood, especially in the country, than a prevailing taste for this pursuit. The freemasonry amongst Archers is proverbial. It has, moreover, one peculiarly bright feature, and this is, that its practice is allowable to both sexes. Hardly any other amusement admits of this. Cricket, tennis, billiards, hunting, &c., are all most excellent, but are confined to the sterner sex, or should be. They have all this failing, that they lack the humanising presence of the fairer portion of humanity. I do not mean as spectators, but as sharers in the exercise. I hope to see the day, when no town or neighbourhood of any extent is without its Archery Club. Each will be the better in many respects for it, without doubt.

And now, before bringing these pages to a conclusion, I must remind my readers of the statement with which they were commenced, namely, that the result of my own practical experience alone would be given. This has been done throughout; and where I have differed most from commonly received rules and practice, *e.g.*, the shape of the bow, the line of pull, the point of aim, &c., I have done so, not from a love of novelty or from a desire to be thought wiser than my fellows, but from positive conviction of the incorrectness of those rules—such conviction being the result, not only of much thought, but of long and patient experiment. I can only say to every brother Archer, with all truth, and a mind open to conviction—

> Si quid novisti rectius istis
> Candidus imperti, si non his utere mecum.

My observations, however, have professed to be addressed more particularly to the young Archer; and to the few words of practical advice ventured upon in my introductory paper, I would now additionally impress upon him the absolute necessity of *perseverance* and a *command of temper*. Without these essentials he will never become eminent as an Archer—neither the idle nor the irritable need hope for success here. This observation, indeed, applies to all pursuits, but is particularly mentioned in connection with Archery, because many (I know several), whilst acknowledging its truth as regards everything else, seem to think that this particular pursuit of Archery should form an exception to the rule, and look upon themselves as positive martyrs because they cannot attain the highest position in a very short time, and without the trouble of working hard for it. This is a great mistake. To obtain a thorough command of the bow is no easy matter; on the contrary, it is a most difficult one; and pre-eminence here requires the exercise of the same qualities as pre-eminence in anything else. So let the young Archer look to it.

One other thing is also particularly desirable for him, though not so absolutely essential, and that is *good teachiug*. The persevering Archer may, and probably will, work his way without it, but only at the cost of an infinity of trouble and disappointment, that otherwise might well be spared him. Faults are more easily fallen into than got rid of afterwards; and this remark is the more applicable to Archery, inasmuch as success in it depends upon the accurate mastery of a number of small difficulties—a failure in any one of them in all probability spoiling the shot. By all means, then, let the beginner, whenever he has the chance, avail himself of superior knowledge and experience. The road to the bull's-eye he will find all the easier by having a finger-post to direct him in the track. He will also find it a good plan to attend a meeting of the Grand National Archery Society as a *spectator*, and, by carefully noting the style, apparatus, and leading points of the best Archers, will be the better able to rectify and guide his own private practice afterwards. Perfection, to imitate, he will find in no single shooter; but, in marking the faults he sees, so as to avoid them, and the excellencies, so as to imitate them, he will find the surest means of approaching it. In short, he who will use his eyes and brains, as well as his hands—who will persevere in the face of failure and disappointment, and not be above taking the best advice, if he chanced to have it offered him, will, in the end, be sure of a very high position.

Having paid so much attention to the beginner, I feel called upon to take one shot at that Archer of long-standing who is very probably giving a sigh of relief as he comes to the concluding chapter of this "tiresome" treatise. Well, Sir, you have practised with the bow, I perceive, some fifteen or twenty years or more, and think, in consequence, you *must* know all about it. You have hitherto differed from me upon most of the essential points treated of in the foregoing pages, and, having since read them, you do so still. *You* will have none of these new-fangled ideas; *you* shoot as your great-great-grandpapa did before you (or was supposed to do —'tis the same thing, of course), and you mean to continue in the same all the days of your life. Even supposing—ridiculous idea!— that the theories propounded and rules laid down are better than your own, still, you say, they will not do for you (at least many of your style of thinking *do* say so)—*you* will have none of them, though those who act upon them prove by results how far they are beyond anything *you* know on the subject. *You* can generally put in one out of three of your arrows at 100 yards, and now and then (twice, perhaps, in three years,) actually accomplish the *half* of them! You can, besides, shoot the other distances in proportion, and *you* are content. Anything beyond these stupendous attainments *must* be owing to luck, or some peculiar gift of nature; for have not you been practising a great number of years earnestly and energetically, and been unable to accomplish more? Since, at any rate, your own ideas *must* be best, so far as you yourself are concerned, a different system, whatever it may do for others, cannot by any possibility, you think, do anything for you.

Now, my very self-opinionated friend, I will venture to reply to you, that no system radically wrong and unscientific in theory, opposed to the plain rules of practical sense, and unsuccessful in its results, can be made right either by its having antiquity pleaded in its favour (supposing you can justly do this, which I very much doubt), nor by any special peculiarities of a particular individual. You *may* have some physical incapability that may prevent your ever becoming a good shot: but if you can use your limbs and muscles like other men, there can be nothing that can render a bad system better for you to act upon than a good one. If you cannot succeed with the latter, you are still less likely to do so with the former. You will by no means get over your unavoidable difficulties the easier by persisting in a mode of shooting that only increases them. I do not presume to say that the system advocated in the foregoing articles is the best that can be discovered; but I say that, so long as it accomplishes in myself and others what the wildest flights of your imagination never dream of as possible to be accomplished by yourself, with your own, it is to be preferred to yours. You may say you have tried it, but find you only shoot the worse. This is likely enough, because no man can change from one mode of doing a thing to another, though it be from a bad to a good one, and not for the time experience an increase of difficulty. The habits of years are not to be so easily overcome. A little perseverance, however, would soon put a different face upon the matter. Of this at any rate be sure, that if you wish to arrive at the top of the tree in any pursuit, you *must* adopt that mode of action that is indisputably proved to be the only, or the best, means of carrying you there.

It would be ungallant, indeed, were I to come to a conclusion without a few words of direct address to the fairer portion of humanity. To you, then, fair Marians, and to you who, though not as yet enrolled in that band, may still, it is hoped, some day be so, let me observe that Archery is a boon indeed. Your sex have few out-door exercises at all—none, with the exception, perhaps, of riding (which is accessible but to few), that at all brings the muscles *generally* into healthy action. You cannot say that mere walking or shop-lounging does this; still less that the heated atmosphere of a ball-room allows of it. But Archery does. How many consumptions, contracted chests, and the like, think you, might have been spared, had its practice been more universal amongst you? It is an exercise admirably suited to meet your requirements—general and equal, without being violent—calling the faculties, both of mind and body, into gentle and healthy play, yet oppressing none—bringing roses to your cheeks, and occupation to your mind, —withal most elegant and graceful. I never took up an Archery book yet, without finding a quotation from one Madame Bola, a celebrated opera dancer, declaring the attitude of the shooter the most graceful in the world: without inflicting the full quotation upon you, suffice it to say that I, in common with every brother Archer, most cordially agree with that most respected female. A "duck" of a bonnet, or of a moiré, has no chance in its killing powers against a "duck" of a shot. Cupid, like Paul Pry, "drops in" as he pleases, anywhere and everywhere; but I think he is particularly fond of a social Archery gathering. Small blame to him for the same! Need I say more to recommend this pastime to you? I think not;—every consideration should induce you to adopt it.

And now, gentle reader, my task is done. To me it has been a labour of love, and to you I trust not altogether useless nor uninteresting. I have, as I proposed, laid before you faithfully and unreservedly, the result of my own experience, and though I have not the vanity to suppose that I have pleased all, I hope that I have offended none. If I have spoken strongly against some errors of opinion or practice, it has been against the error itself, and not against the, perhaps thoughtless, holder of that error.

"Immedicabile vulnus,
Ense recidendum est, ne pars sincera trahatur."

My only aim has been to benefit you; the result is with yourself. Farewell and—shoot well.

Chapter XVI.

ON THE FORMATION, RULES, AND REGULATIONS OF SOCIETIES.

There are certain special requirements applicable to all Societies, of which economy, convenience of place of meeting, and a proper system of prizes are perhaps the most important. Without these, however favorably a Society may commence its career, it will, sooner or later—and much oftener the former than the latter—dwindle away, and come to an untimely end. For if, in the first place, a judicious economy in conducting its affairs be not studied, many will be prevented joining at all, and others be compelled after a time to quit its ranks; for a locality is seldom so wealthy that the larger portion of its society is not limited in its expenditure, in matters of mere amusement. In the second place, should the place of meeting be too distant, or devoid of the necessary accommodation, few will make a point of attending any but the grand prize meetings, neglecting those established for practice, which, after all, conduce the most to the general improvement of the shooting, and to keep alive the spirit of emulation and that social fellowship which ought to prevail amongst the members. In the third place, if there be not such an arrangement of the prizes as that, while merit meets with its due reward, the novice may not be altogether hopeless of success, many a young Archer will be discouraged at the outset, from the knowledge that none but the most skilful have any chance of appearing on the prize-list. Not that I mean to say that the mere money value of any prize is, or should be, the motive to exertion; but it is a fact, amply borne out by experience, that, without the hope of gaining these trophies of success, though they may be of no intrinsic worth, sufficient inducement to the practice necessary for improvement is, in the majority of cases, wanting. With these few preliminary remarks, I will now proceed to the subject of this chapter.

The first step towards the formation of an Archery Society in any locality is, for those who are really ardent in the cause, to meet together as a Provisional Committee, to draw up rules and regulations applicable to the neighbourhood; then to invite, by circular or otherwise, all likely to take an interest in the Society, in order to discuss those rules and regulations so drawn up, and adapt them, as far as may be, to meet the views of all. The following, I conceive, to be such as will in general be found to work the best, but subject, of course, to any modifications called for by the special circumstances of the locality in which the Society is formed :—

RULES.

1.—The Society shall be called . . .

2. The place of meeting shall be . . .
Any person residing within fifteen miles of the same shall be eligible as an actual member; beyond that distance, as an actual or honorary member. No person *resident* within the prescribed limits shall be admitted to any meeting of the Society unless an actual member.

[*If this be thought too stringent a rule, it may be relaxed by allowing invitations to be issued by the Committee. In some Societies the rule is to meet, not at any fixed place, but in private grounds, by invitation; but this cannot be recommended for general adoption, inasmuch as few possessing places suitable for a meeting, the onus falls almost invariably upon those few. Another objection to this custom is, that there is no fixed place of meeting for practice.*]

3.—There shall be a Lady Patroness, a President, Vice-President,

Treasurer, Secretary, and Committee; all but the Treasurer and Secretary to be elected annually, by the members,—these two by the Committee, from their own body.

4.—There shall be an Annual General Meeting of the members of the Society, to audit the accounts of the previous year, elect officers, and transact any other business of the Society. In case of any special emergency, a General Meeting of the Society may be called by the Committee.

5.—The election of new members may take place at any meeting of the Society, provided the name of the party, with his proposer and seconder, shall have been entered in the Secretary's book one month before the election, and that the members be duly notified of the same.

6.—The election shall be by ballot; one black ball in . to exclude; . members to vote, or no election.

[If the system of balloting be objected to, the following rule may be substituted:—the name of the party, with his proposer and seconder, shall be submitted to the Committee, and if approved by them, to the first meeting of the Society; and the candidate declared duly elected, unless a ballot be demanded, when the same shall at once take place. The member desiring the ballot should be allowed to do so privately.]

7. The annual subscription for actual members shall be £1 1*s*. for a lady or gentleman, or £2 2*s*. for a family; only one gentleman above the age of twenty-one years to be included in the family subscription. The entrance fee to be half the subscription.

[These amounts are not arbitrary, but are intended to be either increased or diminished, according to the decision of each individual Society.]

8. The subscriptions shall become due on the .
in each year; such subscriptions to be paid into the hands of the Treasurer as soon as possible after they become due, or at the first meeting. No one to shoot for prizes until the subscription is paid. Any member wishing to withdraw from the Society shall give at least one month's notice to the Secretary previous to the expiration of the current year, otherwise his subscription shall be considered due for the year following.

[*The enforcement of this rule will be found absolutely necessary, to enable the Committee satisfactorily to conduct the financial affairs of the Society.*]

9.—Persons residing beyond the fifteen mile limit, but temporarily staying within it, may be admitted for a shorter period than one year, according to a certain scale of payment to be determined by the Committee.

[*This rule would apply only to certain localities, such as large towns and watering places.*]

10.—At all Meetings the President shall preside, but if absent, the Vice-President. Should neither be present, a Chairman shall be elected for the day, who shall have a casting vote.

11.—No new rule shall be made excepting by the vote of the majority of the members at a General Meeting.

12.—The Secretary shall enter all the proceedings of the Society in a book kept for the purpose.

13.—The costume of the Society shall be . . .

[*I think this is a point too generally neglected, as certainly the variety of inelegant dress now so often seen at an Archery Meeting, more especially amongst the gentlemen, is anything but conducive to a good effect. Not that I would advise anything like an approach to conventional Robinhoodism of costume; but at the same time, it is difficult to associate the idea of Archery with trowsers à la peg-top, a coat a cross between a railway wrapper and a dressing-gown, and a wide-awake, whose original shape it would be difficult even to guess at. I think we may go to the extent of a dark-green coat and cap for the gentlemen, whilst for the ladies, a combination of green and white will always be becoming and elegant. I can only add, that simplicity and economy should go hand in hand.*]

14.—There shall be a pic-nic dinner at each bow meeting, each member or family to bring sufficient for themselves, friends, and servants, also plates, dishes, knives, forks, &c., &c., unless the funds of the Society will admit of the purchase of a sufficient stock of these necessary articles.

[*I have had some hesitation as to this rule, being aware that in the majority of cases it is not adopted; nevertheless, I have found the dissatisfaction and inconvenience caused by other plans so great, as to convince me of its being, on the whole, the best.*]

15.—Dinner shall be served at that hour most convenient to the shooting, and tea and coffee, provided by the Society, immediately after the shooting is over.

16.—No smoking to be allowed on the ground in the presence of the ladies, under a penalty of a fine, to be fixed by the Committee.

17.—All bets to be forfeited to the Club.

18.—The wife and family of any member shall be eligible, without ballot.

19.—A discretionary power shall be given to the Committee to meet any circumstances not contemplated in these rules.

SHOOTING REGULATIONS.

1.—There shall be three prize bow meetings, at least, during the Archery season.

[*These will be probably found sufficient in country clubs, where the members have mostly to come a considerable distance to the place of meeting; but in town clubs, and those established in populous districts, it will be found highly beneficial to have weekly target days also, at which minor honorary prizes may be contended for.*]

2.—The shooting at these prize meetings shall commence at o'clock, and continue until dinner time; after which it shall be resumed, till the appointed number of arrows have been shot; when the scores shall be made up and the prizes awarded.

3.—The prizes shall be awarded as follows:—the first lady's prize for the greatest gross score; the second, for the highest number of hits; the third, for the most central gold; and if there be a fourth, for the most golds;—and the same for the gentlemen.

[*It will add considerably to the amusement of the day if an honorary wooden spoon be given for the lowest gross score, or the greatest number of outer whites.*]

4.—The distances for the ladies shall be 60 yards and 50 yards; and for the gentlemen, 100 yards, 80 yards, and 60 yards; the targets, in each instance, being pitched about three yards further to allow for standing room; their centres shall be exactly four feet from the ground.

5.—The number of arrows shot shall be—for the ladies, at 60 yards, and at 50 yards; and for the gentlemen, at 100 yards, at 80 yards, and at 60 yards.

[*As uniformity is desirable, it would be as well were the*

national round the number fixed on, namely, 48 *at* 60 *yards and* 24 *at* 50 *yards for the ladies; and* 72 *at* 100 *yards,* 48 *at* 80 *yards, and* 24 *at* 60 *yards for the gentlemen,—but this must be determined by the wishes of the majority.*]

6.—Before the shooting commences, a captain shall be appointed at each target, to have the command of it, to superintend the marking, and to see that order and regularity are observed throughout. If any Archer be not ready to shoot in his turn, he shall shoot last, and if not ready then, shall lose that end altogether. The order of shooting at each target shall be according as the names are entered on the scoring card, the captain shooting second.

7.—No one shall shoot out of his turn, or draw an arrow from the target until it has been registered by the marker, under the penalty of loosing its value altogether.

[*This is a very important rule to prevent mistakes or disputes, and should be rigidly enforced by the captain.*]

8.—Each shooter shall be ready in his turn, with his bow strung. Those who have not shot shall stand on a level with, or behind the target, on the left, each advancing in his turn to the mark in its front, shooting his three arrows, and retiring to the right. No Archer or spectator shall move from his place, or cross the ground, until all have shot, or distract the attention of the shooter by remarks, or loud talking.

9.—An arrow breaking two circles shall count for the higher; any dispute on this point shall be decided by the majority of the shooters at that target. The value of the circles shall be 9 for the gold, 7 for the red, 5 for the blue, 3 for the black, and 1 for the white. In settling the total of the scores, the score and hits shall be reckoned separately, and not added together.

[*The plan adopted at the Grand National Meetings, is to add together the relative position of each Archer both in hits and score, and to determine his position by the lowest number. For instance, A. is 1st in score, and 3rd in hits, making a total of 4,—while B. is 1st in hits, and 2nd in score, making a total of 3 only; B. would therefore take precedence of A. by one point.*]

10.—Members who have won prizes for gross score or gross hits, shall be weighted as follows:—Winners of one prize to lose the white circle; of two prizes, the black and white; of three prizes, the blue, black, and white circles, at all distances; but this shall not apply to any challenge or extra prize (unless expressly limited by the donor), nor shall more than the three outer circles be taken away. The loss of circles to be permanent, or until all the members shall be reduced to the red and gold, when they shall commence *de novo*.

11.—In the case of a tie for gross score, the prize shall be awarded to the one having the most hits; in the case of a tie for hits, to the one having the highest score; in the case of a tie both in score and hits, or in the number of golds, or the best gold, it shall be decided by lot.

12.—None but actual members shall receive the Society's prizes.

[*It is usual, however, to have a special prize for honorary members and visitors, when the funds of the Society admit of it.*]

13.—In order to prevent confusion, each member shall have his arrows properly marked, or he shall forfeit any score made with such arrows.

14.—Any person presenting a prize to the Society, shall have full power to regulate the terms on which it shall be awarded.

15.—Every member shall be at liberty to introduce his friends to be present or to shoot at any meeting, subject to rule 2 and regulation 12.

16.—There shall be a challenge prize for the ladies, and another for the gentlemen, to be shot for at each prize bow-meeting, and shall be awarded to . . .

[*This may be either for gross score, or for relative gross score and hits, as explained in note to regulation* 9.]

These prizes shall be the property of the Society, and be held by the winner for the time being, as a mark of superior skill.

17.—An extra prize shall be awarded at the last bow-meeting of the Society to the lady and gentleman making the highest aggregate gross score during all the meetings of the year.

[*This is a very useful prize, as promoting regularity of attendance amongst the members.*]

Printed by H. Davies, Montpellier Library, Cheltenham.